Developing  *izers* with VCV Rack

Developing Virtual Synthesizers with VCV Rack takes the reader step by step through the process of developing synthesizer modules, beginning with the elementary and leading up to more engaging examples. Using the intuitive VCV Rack and its open-source C++ API, this book will guide even the most inexperienced reader to master efficient DSP coding to create oscillators, filters, and complex modules.

Examining practical topics related to releasing plugins and managing complex graphical user interaction, with an intuitive study of signal processing theory specifically tailored for sound synthesis and virtual analog, this book covers everything from theory to practice. With exercises and example patches in each chapter, the reader will build a library of synthesizer modules that they can modify and expand.

Supplemented by a companion website, this book is recommended reading for undergraduate and postgraduate students of audio engineering, music technology, computer science, electronics, and related courses; audio coding and do-it-yourself enthusiasts; and professionals looking for a quick guide to VCV Rack. VCV Rack is a free and open-source software available online.

Leonardo Gabrielli, PhD, is a research fellow at the Department of Information Engineering, Università Politecnica delle Marche, Italy, where he lectures on music production and conducts research on physical modeling and deep neural audio processing. He collaborates with the music industry as a DSP developer and he is scientific director of Acusmatiq MATME.

Developing Virtual Synthesizers with VCV Rack is a great read; an informative read; an inspiring read; an empowering read; an essential read! This fantastic textbook and comprehensive set of applied audio coding tutorials by Gabrielli will become the basis of many college courses and foundational courses of study for every electronic musician interested in knowing how things work under the hood. With that knowledge and understanding, they can code their own software synthesizers and DSP modules in VCV Virtual Eurorack format.

Dr. Richard Boulanger, Professor of Electronic Production and Design,
Berklee College of Music

If, like me, you enjoy creating your own music tools, this book will help you bring your ideas to life inside the wonderful world of VCV Rack. Whether you are a beginner or an expert in DSP, you will learn something interesting or new.

Dr. Leonardo Laguna Ruiz, Vult DSP

Developing Virtual Synthesizers with VCV Rack

Leonardo Gabrielli

Routledge
Taylor & Francis Group

NEW YORK AND LONDON

First published 2020
by Routledge
52 Vanderbilt Avenue, New York, NY 10017

and by Routledge
2 Park Square, Milton Park, Abingdon, Oxon, OX14 4RN

Routledge is an imprint of the Taylor & Francis Group, an informa business

© 2020 Taylor & Francis

The right of Leonardo Gabrielli to be identified as author of this work has been asserted by him in accordance with sections 77 and 78 of the Copyright, Designs and Patents Act 1988.

Library of Congress Cataloging-in-Publication Data
Names: Gabrielli, Leonardo, author.
Title: Developing virtual synthesizers with VCV Rack / Leonardo Gabrielli.
Description: New York, NY : Routledge, 2020. | Includes bibliographical references and index.
Identifiers: LCCN 2019039167 (print) | LCCN 2019039168 (ebook) | ISBN 9780367077730 (paperback) | ISBN 9780367077747 (hardback) | ISBN 9780429022760 (ebook) | ISBN 9780429663321 (epub) | ISBN 9780429660603 (mobi) | ISBN 9780429666049 (adobe pdf)
Subjects: LCSH: VCV Rack. | Software synthesizers. | Computer sound processing. | Sound studios–Equipment and supplies.
Classification: LCC ML74.4.V35 G325 2020 (print) | LCC ML74.4.V35 (ebook) | DDC 786.7/4134–dc23
LC record available at https://lccn.loc.gov/2019039167
LC ebook record available at https://lccn.loc.gov/2019039168

ISBN: 978-0-367-07774-7 (hbk)
ISBN: 978-0-367-07773-0 (pbk)
ISBN: 978-0-429-02276-0 (ebk)

Typeset in Times New Roman
by Swales & Willis, Exeter, Devon, UK

Visit the companion website: www.leonardo-gabrielli.info/vcv-book
Lightbulb icon by Pettycon from Pixabay

Printed and bound by CPI Group (UK) Ltd, Croydon, CR0 4YY

Contents

Preface

In September 2017, a new open-source software project was released in the wild of the Internet, starting from a few hundred lines of source code written by a talented developer. A couple of months later, thousands of passionate musicians and developers were eagerly following its quick growth. With a user base of tens of thousands, VCV Rack is now one of the trendiest platforms for modular software synthesis, and it is particularly appealing for its open-source codebase and ease of use.

In October 2017, I was more than happily having fun out in the sun during the weekends, trying to soak up the last sun of the trailing summer. My teaching and R&D activities were already quite engaging and kept me more than busy. However, after a glimpse at its source code, I was instantly aware of the potential of this platform: VCV Rack was the first platform ever that would really allow coders, even self-taught ones, to build sounding objects at ease. The modern and swift way of programming a plugin allows almost anyone with some coding background to create their own modules. The sample-wise processing architecture allows students to focus on signal processing experiments rather than spending time debugging cumbersome buffering mechanisms or learning complex application programming interfaces (APIs). As a teaching tool, it allows students to learn real-world coding instead of plugging graphical blocks or writing scripts in scientific computing platforms. As a prototyping tool, it also allows developers to test DSP code quickly before adapting it to their own framework and delivering it.

I was really tempted to help spread the use of this platform and write a teaching resource for students and enthusiasts. Encouraged by good friends and synthesizer experts, I got into this adventure that eventually demanded quite a lot of my time. I hope this resource will prove useful for beginners and students, allowing them a quick bootstrap into the world of virtual synthesizers, sound synthesis, and digital signal processing (DSP), and I really believe that Rack has the potential to serve as a great didactical tool in sound and music engineering teaching programmes.

Finally, I humbly apologize for mistakes or shortcomings that may still be present in the book. The preparation of this manuscript required a lot of effort to track the swift software changes occurring in the beta versions, before Rack v1 would freeze, and to release a timely resource in this liquid world required me to go straight to the point and prepare a compact book that could serve as a primer, an appetizer, or a quick guide, depending on the reader's needs. Arranging the contents was not an easy task either. Writing the theoretical section required a lot of trade-offs to adapt to heterogeneous readers. I hope that my take on DSP and the selection of modules will make most of the readers happy.

Objectives

This book is meant as a guide to get people with some object-oriented programming basics into the development of synthesizers. Some DSP theory is covered, specifically tailored for sound synthesis and virtual analog, sufficiently intuitive and suitable for readers not coming from a technical background (some high school math is required, though). There are many books covering sound synthesis and DSP for audio, with examples using scientific programming languages, quick to use, but far from a real-world implementation. Two books published by Focal Press provide a thorough guide for the development of audio effects and synthesizer plugins in C++. I regard my book as the perfect starting point. It covers the foundations of digital musical instruments development and prototyping, gets the reader involved with sound synthesis, and provides some theory in an intuitive way. It is also an important resource to quickly learn how to build plugins with VCV Rack and get into its specificities. After this, one could move to more complex platforms such as the VST/AU software development kits (SDKs) or get involved with more theoretical studies. The book is also an important resource to quickly learn how to build plugins with VCV Rack and get into its specificities.

The VCV Rack API is very simple, and it only takes a few minutes to create a module from scratch. Unlike other commercial SDKs, you will not lose your focus from what's important: learn the DSP foundations, improve the sound quality of your algorithms, and make the user experience immediate. You will get experienced with algorithms and be able to prototype novel modules very easily. The API is general enough to allow you get your code to other platforms if you wish. But if you want to get some revenue from your coding, there is no need to shift to other commercial SDKs, as the VCV Rack plugin store has numerous options for selling your products. Experienced developers will thus find sections related to the release of commercial plugins.

Reading the book

The guide will provide some basic math and will stimulate intuition to help the reader develop an understanding of the DSP theory that is the common ground to all digital audio algorithms. Most of the theory will be covered in Chapter 2, after a first introduction to modular synthesizers, covered in Chapter 1.

A step-by-step guide to compiling the software and building very simple plugins is provided in the following chapters. We raise the bar, progressively, by introducing selected foundational concepts of synthesizers, virtual analog, and DSP. These will not be presented in an orthodox, academic way, but in a more intuitive way, to facilitate comprehension to any reader, especially the novice ones.

A prerequisite for following the book is that the reader knows a bit about object-oriented programming (OOP) in C++ or similar languages. For readers coming from coding experiences with Arduino and such, the mind framework must be adjusted from sequential coding to OOP. Readers with a background in C and experience with platforms such as Csound and Pure Data will find this platform easier to work with. I suggest to all readers wishing to improve their coding skills to avoid copy-pasting snippets of code, but rather get a good C++ text to refine one's skills with OOP and C++. The learning curve is steeper, but in the long run you will be rewarded. The best professionals are those who seriously spent time on studying, and the world is full of copy-pasters anyway.

Chapters 3 and 4 discuss the platform and how to start building Rack and its plugins. Chapter 5 provides a first introduction to the design of the graphical user interface for Rack modules, while later in Chapter 9 more advanced concepts are provided. Chapters 6, 7, and 8 introduce readers to the development of plugins by proposing several examples of increasing mathematical complexity. Finally, Chapter 10 deals with a few extra topics of general interest regarding the development process, how to get into the third-party plugin list, and how to make some money out of your modules. A final chapter will suggest some ideas for further reading and new projects.

This book details the development of several synthesizer modules. These are collected in a Rack plugin called ABC, available online at www.leonardo-gabrielli.info/vcv-book. Each module is meant to teach something new about Rack and its API; look out for the tips, indicated by a light bulb icon. The theoretical section has been enriched with boxes where some intuitive hints are given and mappings between theory and practice are drawn to improve readability and stimulate the reader. As an engineering student, I always made myself mappings between theoretical concepts and audio systems; this helped me a lot to remember, understand and have fun. Exercises are also available, proposing little challenges to start working autonomously. Some of the online modules include solutions for the exercises, highlighted using C preprocessor macros.

The ABC modules are designed to be simple to avoid confusing the reader with extra functionalities. They are not fitted with lots of bells and whistles, and they are not meant to be the best in their category; they are just meant to get you straight to the point and gain an extra piece of information each time. The plugin store is full of great plugins, and it is up to you to challenge other developers and come up with even better ones. For this reason, all ABC plugins humbly start with an "A." I describe a sequencer "ASequencer," a clock "AClock," and so forth. Maybe you will be the one to develop "the" sequencer and "the" clock that anybody will use in the Rack community! Have fun!

Acknowledgments

Lots of people have given their contribution to this book during its conception and writing.

First of all, the book would have never been started without the enthusiastic support of Enrico Cosimi, who is a fabulous teacher, a charismatic performer, and a great entertainer.

The work has drained much of my energies, patiently restored by Alessandra Cardinali. She also provided useful suggestions on the math and frequent occasions of discussion.

The mechanical design and the crafting of metal parts for the hardware Rack prototype has been carried out with the help of Francesco Trivilino, who is also a valuable modular musician and constructor of rare machines, such as reproductions of Luigi Russolo's Intonarumori.

I'm thankful to Stefano D'Angelo for suggesting some corrections and to Andrew Belt for providing insights. Reading his code is a breeze and there is a lot to learn from him.

My involvement with modular synthesis would have never been so strong without ASMOC (Acusmatiq-Soundmachines Modular Circus), a yearly performance format that was created by Paolo Bragaglia and Davide Mancini after a boastful brainstorming that included Fabrizio Teodosi.

Finally, I want to state my sincere gratitude to my family for transmitting my love for my region, Le Marche, a land of organic farmers and ill-minded synth manufacturers. My love for musical instrument crafting is intertwined with my connection to these towns and the seaside. Thanks to my dad for passing on my the love for musical instrument design, coding, and signal processing. I feel like a third-generation music technologist after him, and my grandfather, who designed the first plastic reeds for melodicas. I hope to keep the tradition alive and take it to the next level, with emerging technologies and paradigms, and I hope that this book will help lots of students and enthusiasts in getting engaged with digital instrument design and development.

CHAPTER 1

Modular Synthesis: Theory

The scope of this book is twofold: while focusing on modular synthesizers – a very fascinating and active topic – it tries to bootstrap the reader into the broader music-related DSP coding, without the complexity introduced by popular DAW plugin formats. As such, an introduction to modular synthesis cannot be neglected, since Rack is heavily based on the modular synthesis paradigm. More specifically, it faithfully emulates Eurorack mechanical and electric standards. Rack, as the name tells, opens up as an empty rack where the user can place modules. Modules are the basic building blocks that provide all sorts of functionalities. The power of the modular paradigm comes from the cooperation of small, simple units. Indeed, modules are interconnected at will by cables that transmit signals from one to another.

Although most common hardware modules are analog electronic devices, I encourage the reader to remove any preconception about analog and digital. These are just two domains where differential equations can be implemented to produce or affect sound. With Rack, you can add software modules to a hardware setup (Section 11.3 will tell you how), emulate complex analog systems (pointers to key articles and books will be provided in Section 11.1), or implement state-of-the-art numerical algorithms of any sort, such as discrete wavelet transform, non-negative matrix factorization, and whatnot (Section 11.2 will give you some ideas). To keep it simple, this book will mainly cover topics related to oscillators, filters, envelopes, and sequencers, including some easy virtual analog algorithms, and provide some hints to push your plugins further.

1.1 Why Modular Synthesis?

Why do we need modularity for sound generation?

Most classical and contemporary music is structured using modules, motives, and patterns. Most generating devices can be divided into modules either physically or conceptually. I know, you cannot split a violin in pieces and expect them to emit sound singularly. But you can still analytically divide the string, the bridge, and the resonant body, and generate sound by emulating their individual properties and the interconnection of the three.

Modularity is the direct consequence of analytical thinking. Describing the whole by dividing it into simpler components is a strategy adopted in most scientific and engineering areas. Unfortunately, studying the interconnection between the parts is often left "for future works." Sometimes it does no harm, but sometimes it leaves out the largest part of the issue.

I like to think about the modular sound generation paradigm as the definitive playground to learn about the concept of separation and unity and about the complexity of nonlinear dynamical systems.

Even though each module in its own right is well understood by its engineer or developer, the whole system often fails to be analytically tractable, especially when feedbacks are employed. This is where the fun starts for us humans (a little bit less fun if we are also in charge of analyzing it using mathematical tools).

1.2 An Historical Perspective

1.2.1 The Early Electronic and Electroacoustic Music Studios

As electronic music emerged in the twentieth century, a large part of the experimental process involved with it is related to the first available electronic devices, which were often built as rackmount panels or heavy cabinets of metal and wood. The 1950s are probably the turning point decade in the making of electronic music. Theoretical bases had been set in the first half of the century by composers and engineers. The invention of the positive feedback oscillator (using vacuum tubes) dates back to 1912–1914 (done independently by several researchers and engineers), while the mathematical formalization of a stability criterion for feedback systems was later devised by Heinrich Barkhausen in 1921. The field-effect transistor was patented in 1925 (in Canada, later patented in the US as patent no. 1,745,175), although yet unfeasible for its manufacturing complexity. After World War II, engineering had evolved wildly in several fields, following, alas, the war's technological investment, with offspring such as control theory and cybernetics. Purely electronic musical instruments were already available far before the end of the war (e.g. the Theremin in 1920 and the Ondes Martenot in 1928), but still relegated to the role of classical musical instruments. Noise had been adopted as a key concept in music since 1913 with Luigi Russolo's futurist manifest *L'arte dei rumori*, and atonality had been developed by composers such as Arnold Schoenberg and Anton Webern. In the aftermath of World War II, an evolution was ready to start.

A part of this evolution was driven by European electronic and electroacoustic music studios, most notably the WDR (Westdeutscher Rundfunk) in Cologne, the RTF (Radiodiffusion-Télévision Française) studio in Paris, and the *Centro di Fonologia* at the RAI (Radiotelevisione Italiana) studios in Milan.

The Cologne studio was born in the years 1951–1953. Werner Meyer-Eppler was among the founders, and he brought his expertise as a lecturer at the University of Bonn in electronic sound production (Elektrische Klangerzeugung) into the studio. Karleinz Stockhausen was involved from 1953, playing a key role in the development of the studio. He opposed to the use of keyboard instruments of the time (e.g. the Melochord[1] and the Monochord, introduced in the studio by Meyer-Eppler) and turned to a technician of the broadcast facility, Fritz Enkel, for getting simple electronic devices such as sine wave generators.

The Paris studio was employed by the *Groupe de Recherches Musicales*, which featured pioneers Pierre Schaeffer, Pierre Henry, and Jacques Poullin, and was mostly based on Schaeffer's concepts of concrete music, and therefore mostly oriented to tape works and electroacoustic works.

Finally, the studio in Milan was born officially in 1955 and run by composers Luciano Berio and Bruno Maderna and technician Marino Zuccheri. Most of the equipment was designed and assembled by Dr. Alfredo Lietti (1919–1998), depicted in Figure 1.1 in front of his creatures. It is interesting to note that Lietti started his career as a radio communication technician, showing again

Figure 1.1: C.1960, Alfredo Lietti at the Studio di Fonologia Musicale, RAI, Milan. Lietti is shown standing in front of rackmount devices he developed for the studio, mostly in the years 1955–1956. A cord patch bay is clearly visible on the right. *Credits*: Archivio NoMus, Fondo Lietti.

how communication technology had an impact on the development of electronic music devices. The Studio di Fonologia at its best boasted third-octave and octave filter banks, other bandpass filters, noise and tone generators, ring modulators, amplitude modulators, a frequency shifter, an echo chamber, a plate reverb, a mix desk, tape recorders, various other devices, and the famous nine sine oscillators (Novati and Dack, 2012). The nine oscillators are often mentioned as an example of how so few and simple devices were at the heart of a revolutionary musical practice (Donati and Paccetti, 2002), with the stigmatic statement "Avevamo nove oscillatori," *we only had nine oscillators*.

All the electronic music centers of the 1950s had rackmount oscillators, filter banks, modulators, and tape devices. These would be poor means by today's standards to shape and arrange sound, yet they were exactly what composers required at the time to rethink musical creation. Reorganizing musical knowledge required elementary tools that would allow the composers to treat duration, dynamics, pitch, and timbre in analytical terms. The electronic devices used in these early electronic music studies were typically adapted from communication technology.[2] After all, these first studios were hosted by broadcast companies.

What was extraordinary about the first electronic and electroacoustic works of the 1950s was their eclecticism. It turned out that a few rudimentary devices were very flexible and could serve the grand ideas of experimental composers. The research directions that were suggested by early electronic and electroacoustic music composers of the time, such as Olivier Messiaen, Pierre Schaeffer, and Karleinz Stockhausen, proved fruitful. In the works of these studios, we can see the premises of the advent of modular synthesis, where electronic tools are employed creatively, suggesting an idea of the studio as a musical instrument or device. A few fundamental steps were still missing at the time:

- Packing a studio in a case (or in a laptop, as we do nowadays) was impossible. In the 1950s, the technology was yet underdeveloped in terms of integration, size, and power consumption. The first bipolar junction transistor had just been patented and produced in 1948 and 1950, respectively, at the Bell Labs, USA, and years of engineering were required to make production feasible on a large scale and with small form factors.
- Electronic music was very academic and far from audiences, except from early science fiction movies. These new forms of music would still take a couple of decades to leak into popular music, as we shall see in the next section.

1.2.2 The Birth and Evolution of Modular Synthesizers

In the US, things were moving fast from the technical side. In 1951, entertainment devices manufacturer RCA started a research project that led to the development of the RCA Sound Synthesizers Mark I and Mark II, hosted at Columbia University's Computer Music Center. These were modular systems that could be patched by wires and follow a score programmed through punched paper, finally leading to a recording on a lacquer disc (or later to a magnetic tape). They were totally based on vacuum tubes and took up the space of an office. Serial music composers were interested in their capabilities and music programming flexibility, but they never had an impact in contemporary music, with Princeton University's composer Milton Babbit being one of the few prominent users.

In those years, computer music centers were rising in the US. The first and major innovator in the field was Max Mathews. After taking its degree at the Massachusetts Institute of Technology in 1954, Max Mathews started working at Bell Labs in the US, where he devised – in a few years – most of the key concepts of computer music and digital sound synthesis through the development of MUSIC I and later versions. He is still regarded as one of the main innovators in the field. The concepts discussed in Section 4.1 were mostly there in Mathews' works by the end of the 1950s.

The US, however, was also appealing to skilled inventors who sought commercial success and modern manufacturing facilities. German engineer Harald Bode, the creator of the aforementioned Melochord, moved to the US in 1954 to continue developing his ideas. He was among the first ones to foresee transistor technology as a key changer in designing synthesizers and the first to build a compact modular synthesizer. In his 1984 retrospective survey on musical effects, he states that "in the Melochord for the Stockhausen Studio in Cologne, the modular concept was adopted, by which external ring modulators, echo chambers, and the like could be included in the system" (Bode, 1984, p. 732). He also built a modular sound modification system between 1959 and 1960 (Bode, 1961), which included:

the multiplier-type ring modulator and other sound modifying devices [...] such as an envelope follower, a tone-burst-responsive envelope generator, a voltage-controlled amplifier, formant and other filters, mixers, a pitch extractor, a comparator and frequency divider for the extracted pitch, and a tape loop repeater with dual channel processing. The modular concept proved attractive due to its versatility, and it was adopted by Robert Moog when he created his modular synthesizer in 1964.

(p. 733)

Robert Moog (1934–2005), who probably requires no introduction to any readers, started developing modular synthesizers in the 1960s. This eventually led him to assemble the first integrated synthesizer ever based on subtractive synthesis. He presented his first voltage-controlled synthesizer using a keyboard controller at the 16th annual fall convention of the Audio Engineering Society in 1964. The full paper can be found in the *Journal of the Audio Engineering Society* (Moog, 1965). This describes his development of an electronic music system for real-time performance, based on the concepts of modularity, voltage-controlled devices (filters, oscillators, etc.), and the keyboard as a control device. Indeed, the paper shows a prototype consisting of a five-octave keyboard and a small wooden case with wired modules. A part of the paper is devoted to discussing the voltage control paradigm (Figure 1.2).

Robert Moog is best known as the creator of the Minimoog Model D (1970), acknowledged as the first and most influential synthesizer of all time because of its playability and compactness. However, all the prototyping and improvements that made it great would not have been possible

Figure 1.2: Robert Moog's experience with modular systems led him to create the first compact synthesizer, the Minimoog Model D. In this picture, an early prototype system is shown from a scientific paper by Robert Moog. Image courtesy of the Audio Engineering Society (www.aes.org/e-lib/browse.cfm? elib=1204).

without the large amount of work conducted on his modular systems side by side with great musicians of his time.[3] The Model D came after several revisions, and the first of them – the Model A – was composed of modules. All the above shows how modular synthesis, although obscure and underground (until recently), played a key role in the birth of the electronic synthesizer. It is only after years of experimentation and engineering that Moog came out with what would be for years regarded as the perfect companion to keyboard players of the progressive rock era. The Moog Modular and the Minimoog had a tremendous impact on popular music. Stanley Kubrick's *A Clockwork Orange* (1971) featured Wendy Carlos' transpositions of Elgar, Purcell, and Beethoven, as well as other unpublished works of classical inspiration, all played on the Moog Modular. At the same time, jazz and progressive rock keyboard players Sun Ra (1914–1993), Keith Emerson (1944–2016), and Rick Wakeman (1949–) adopted the Minimoog for live usage, inspiring dozens of keyboard players of their time.

While the subtractive synthesizer with keyboard was appealing to many keyboard players, a different approach was carried out by a slightly lesser-known pioneer in electronic music history: Donald "Don" Buchla (1937–2016). He designed modular instruments for composers who were interested in programmable electronic instruments, and his efforts started independently from Robert Moog. During his whole life, Don Buchla fostered the development of modular synthesis, following an experimental approach where user interaction is done through sequencers, touch plates, and the panel itself. Notable composers working with Don Buchla systems were Pauline Oliveros (1932–2016), Morton Subotnick (1933–), Suzanne Ciani (1946–), and Ramon Sender (1934–).

Together with Don Buchla, another famous name in modular synthesizer history is that of Serge Tcherepnin. They are both recognized as references for the so-called *West Coast* approach to synthesis, in opposition to the *East Coast* approach.

1.2.3 The Advent of a Standard

Until recently, modular synthesizers have never been very popular, due to their cost, complexity, and incompatibility between different standards. These differ in the mechanical specifications (size, screws, etc.), the electrical specifications (voltage ranges, pitch voltage values, etc.), and the cable connectors (size, tip-sleeve versus banana, etc.).

A key factor in the resurgence of modular synthesizers is the wide acceptance of a de facto standard, the Eurorack format, developed by a German manufacturer in the late 1990s. Nowadays, most modules are produced according to this format, and major synthesizer manufacturers from Japan and the US are now selling Eurorack-format modules. Nowadays, Eurorack-format modules are getting affordable due to competition, the development of cheap manufacturing facilities, and the sharing of open schematics and knowledge.

The specifications of the Eurorack format follow:

- The rack size follows the 19{DP} rack system (DIN 41494/IEC 297-3/IEEE 1001.1). A rack is 3 height units (HUs) high (133.4 mm) and modules must be large of an integer multiple of 1 horizontal pitch (HP), which is 5.08 mm.
- The power supply is {PM}12 V with an additional 5 V supply.
- Audio signals are usually between {PM}5 V (10 V peak-to-peak, i.e. 10 Vpp).
- Control voltages are bipolar or unipolar 5 Vpp.
- Trigger, gate, and clock signals are unipolar 5 V pulses.

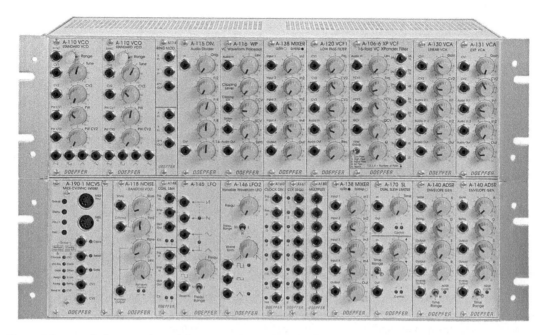

Figure 1.3: A typical Eurorack modular system by Doepfer, including oscillators, filters, envelope generators, and various utility modules. Image courtesy of Doepfer Musikelektronik GmbH.

The Eurorack standard is so widespread that VCV Rack adopts it officially: even though voltages and modules are virtual, all Rack modules comply (ideally) to the Eurorack format. Thus, you should bear these specifications in mind.

In recent years, the arsenal of Eurorack modules has grown, including digital modules (devoted to sequencing, signal processing, etc.), programmable modules, often based on open-source hardware and software (e.g. Arduino and related), or analog-to-digital and digital-to-analog converters for MIDI conversion or to provide DC-coupled control voltages to a personal computer (Figure 1.3).

1.2.4 The Software Shift

The modular boom was followed by the production of software tools. Although software modularity and patching are found in early computer music software such as Pure Data and Max (to name a few), several virtual analog modular emulators have been proposed in the years since the inception of virtual analog technology (1994 onward).

One well-known hybrid system was the Nord Modular, a hardware DSP-based machine that would run modular patches created using a PC editor. The first product of the series was out in 1998 and the last was discontinued in 2009.

Software company Softube sells Modular, a software plugin announced in 2015 and currently sold under a commercial license. It provides emulation of Eurorack, Buchla, and other modules.

Native Instruments' Reaktor is also an alternative software solution for the emulation of modular synthesizers. Despite being a modular software environment since its inception, only with version 6 (2015) did it feature *Blocks* (i.e. rackmount-style modules). One of the highlights in their advertisement is the excellence of their virtual analog signal processing. I should also mention that a lot of research papers in the field are done by employees of this company, and that one of them shares for free via their website the largest reference on virtual analog filters (Zavalishin, 2018), a mandatory read for those interested in virtual analog!

Finally, there's VCV Rack. Launched in September 2017 by Andrew Belt, it is the latest platform for software modular synthesis, and it has grown quickly thanks to its community of developers and enthusiasts. It differs from the commercial solutions above in that it provides an open-source SDK to create C++ modules. But of course, you already know this!

1.3 Modular Synthesis Basics

Thus far, we have not started getting into the technical details of a modular setup. What kind of modules exist? What is the typical setup? How do we create a basic wish list of fundamental modules?

There are no simple answers to these questions. The modular universe is so big, bizarre, and complex that there is no typical setup, no rules, and infinite paradigms. But in trying to keep things practical, a few distinctions and generalizations can be done, of course.

In the Eurorack standard, a signal is just a voltage. A voltage is the electric potential difference between two conductors. As such, all Eurorack signals need two conductors to carry a signal, or, practically speaking, a male 3.5 mm TS (tip-sleeve) jack connector, what you would call a mono minijack. The reference connector connected to ground is the sleeve of the jack, while the proper signal is conveyed by the tip. Modules have input and output female connectors, sometimes called inlets and outlets throughout the book. The output of a module should be connected to the input of another, as shown in Figure 1.4. Connections between inputs and outputs should be avoided.

Now let us move to the modules. There are hundreds of modules in the Eurorack world alone. Most modules, fortunately, can be included in a specific category. Let us describe the most important families of modules.

1.3.1 Sound Sources

If we want to play some music, sound has to be generated somehow, right? Then the first family of modules that we are going to tackle is that of the *sound sources*: oscillators, noise sources, and samplers, mainly. Oscillators are those modules that generate a pitched tone. Their frequency content varies depending on the waveform that is generated. Historically, only simple waveforms were generated, according to the electronic knowledge available. Typical waveforms are thus triangular, rectangular, sawtooth, and sinusoidal. These are all very simple waveforms that can be obtained by a few discrete components. From simple designs come simple spectra: their shape is very straight and unnatural, thus requiring additional processing to obtain pleasant sounds. Their spectral properties are discussed in Section 2.8, after the basic concepts related to frequency-domain analysis have been discussed (Figure 1.5).

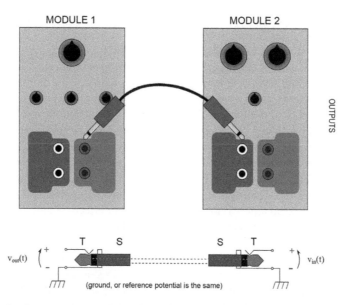

Figure 1.4: Connection between two modules. The output of Module 1 is connected to the input of Module 2 by a TS jack. This way, the input voltage of Module 2 follows the output voltage of Module 1, and thus the signal is conveyed from Module 1 to Module 2.

Figure 1.5: Typical synthesizer oscillator waveforms include (from left to right) sawtooth, triangular, and rectangular shapes.

Oscillators usually have at least one controllable parameter: the pitch (i.e. the fundamental frequency they emit). Oscillators also offer control over some spectral properties. For example, rectangular waveform oscillators may allow pulse width modulation (PWM) (i.e. changing the duty cycle Δ, discussed later). Another important feature of oscillators is the synchronization to another signal. Synchronization to an external input (a master oscillator) is available on many oscillator designs. So-called *hard sync* allows an external rising edge to reset the waveform of the slave oscillator and is a very popular effect to apply to oscillators. The reset implies a sudden transient in the waveform that alters the spectrum, introducing high-frequency content. Other effects known as weak sync and soft sync have different implementations. Generally, with soft sync, the oscillator reverses direction at the rising edge of the external signal. Finally, weak sync is similar to hard sync, but the reset is applied only if the waveform is close to the beginning or ending of its natural cycle. It must be noted, however, that there is no consensus on the use of the last two terms, and different synthesizers have different behaviors. All these synchronization effects require a different period between slave and master. More complex oscillators have other ways to alter the spectrum of a simple waveform (e.g. by using waveshaping). Since there are specific modules that perform waveshaping, we shall discuss them later. Oscillators may allow frequency modulation (i.e. roughly speaking, controlling the pitch with a high-frequency signal). Frequency modulation is the basis for FM synthesis techniques, and can be either linear or logarithmic (linear FM is the preferred one for timbre sculpting following the path traced by John Chowning and Yamaha DX7's sound designers).

To conclude, tone generation may be obtained from modules not originally conceived for this aim, such as an envelope generator (discussed later) triggered with extremely high frequency.

Noise sources also belong to the tone generators family. These have no pitch, since noise is a broadband signal, but may allow the selection of the noise coloration (i.e. the slope of the *spectral rolloff*), something we shall discuss in the next chapter. Noise sources are very useful to create percussive sounds, to create drones, or to add character to pitched sounds.

Finally, the recent introduction of digital modules allows for samplers to be housed in a Eurorack module. Samplers are usually capable of recording tones from an input or to recall recordings from a memory (e.g. an SD card) and trigger their playback. Other all-in-one modules are available that provide advanced tone generation techniques, such as modal synthesis, FM synthesis, formant synthesis, and so on. These are also based on digital architecture with powerful microcontrollers or digital signal processors (DSPs).

1.3.2 Timbre Modification and Spectral Processing

As discussed, most tone generators produce very static sounds that need to be colored, altered, or emphasized. *Timbre modification* modules can be divided into at least four classes: filters, waveshapers, modulation effects, and vocoders.

Filtering devices are well known to engineers and have played a major role in electrical and communication engineering since the inception of these two fields. They are devices that operate in the frequency domain to attenuate or boost certain frequency components. Common filters are the low-pass, band-pass, and high-pass type. Important filters in equalization applications are the peak, notch, and shelving filters. Musical filters are rarely discussed in engineering textbooks, since engineering requirements are different from musical requirements. Among these, we have a low implementation cost, predetermined spectral roll-off (e.g. 12 or 24 dB/oct), and the possibility to introduce a resonance at the cutoff frequency, eventually leading to self-sustained oscillation.[4]

While engineering textbooks consider filters as linear devices, most analog musical filters can be operated in a way that leads to nonlinear behavior, requiring specific knowledge to model them in the digital domain.

Waveshaping devices have been extensively adopted by synthesizer developers such as Don Buchla and others in the West Coast tradition to create distinctive sound palettes. A waveshaper introduces new spectral components by distorting the waveform in the time domain. A common form of waveshaper is the foldback circuit, which wraps the signal over a desired threshold. Other processing circuits that are common with guitar players are distortion and clipping circuits. Waveshaping in the digital domain requires a lot of attention in order to reduce undesired artifacts (aliasing).

Other effects used in modular synthesizers are so-called modulation effects, most of which are based on delay lines: chorus, phaser, flanger, echo and delay, reverb, etc. Effects can be of any sort and are not limited to spectral processing or coloration, so the list can go on.

Vocoders have had a large impact in the history of electronic music and its contaminations. They also played a major role in the movie industry to shape robot voices. Several variations exist; however, the main idea behind it is to modify the spectrum of a first sound source with a second one that provides spectral information. An example is the use of a human voice to shape a synthesized tone, giving it a speech-like character. This configuration is very popular.

Figure 1.6: A CRB Voco-Strings, exposed at the temporary Museum of the Italian Synthesizer in 2018, in Macerata, Italy. This keyboard was manufactured in 1979–1982. It was a string machine with vocoder and chorus, designed and produced not more than 3 km from where I wrote most of this book. Photo courtesy of Acusmatiq MATME. Owner: Riccardo Pietroni.

1.3.3 Envelope, Dynamics, Articulation

Another notable family of effects includes all the *envelope, dynamics,* and *articulation* devices. Voltage-controlled amplifiers (VCAs) are meant to apply a time-varying gain to a signal in order to shape its amplitude in time and create a dynamic contour. They can be controlled by a high-frequency signal, introducing amplitude modulation (AM), but more often they are controlled by envelope generators (EGs). These are tools that respond to a trigger or gate signal to generate a voltage that rises and decays, determining the temporal evolution of a note or any other musical event. Usually, such evolution is described by four parameters: the attack, decay, and release times and the sustain level, producing an ADSR scheme, depicted in Figure 1.7. Most envelope generation schemes follow the so-called ADSR scheme, where a tone is divided into three phases, requiring four parameters:

- *A: The attack time.* This parameter is expressed as a time parameter in [s] or [ms] or a percentage of a maximum attack time (i.e. 1–100).
- *D: The decay time.* The time to reach a steady-state level (usually the sustain, or zero when no sustain is provided by the EG), also expressed as a time ([s], [ms]) or a percentage of a maximum decay time (1–100).
- *S: The sustain level.* The steady-state level to be reached when the decay phase ends. This is usually expressed as a percentage of the peak level that is reached in the attack phase (1–100).
- *R: The release time.* The time to reach zero after the musical event ends (e.g. note off event). This is also expressed in [s], [ms], or percentage of a maximum release time (1–100).

Subsets of this scheme, such as AR, with no sustain phase, can still be obtained by ADSR. An EG generates an envelope signal, which is used as an operand in a product with the actual signal to shape. It is important to distinguish between an EG and a VCA; however, sometimes both functionalities are comprised in one device or module.

Envelope generators are also used to control other aspects of sound production, from the pitch of the oscillator to the cutoff of a filter (Figure 1.8).

Similarly, low-frequency oscillators (LFOs) are used to control any of these parameters. LFOs are very similar to oscillators, but with a frequency of oscillation that sits below the audible range or slightly overlapping with its lower part. They are used to modulate other parameters. If they modulate the pitch of an oscillator, they are performing vibrato. If they modulate the

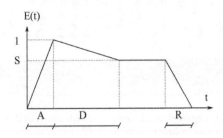

Figure 1.7: A linear envelope generated according to the ADSR scheme.

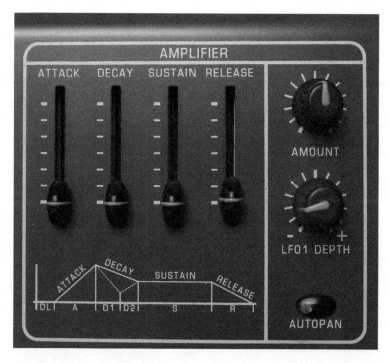

Figure 1.8: Advanced envelope generation schemes may go beyond the ADSR scheme. The panel of a Viscount-Oberheim OB12 is shown, featuring an initial delay (DL) and a double decay (D1, D2) in addition to the usual controls.

amplitude of a tone through a VCA, they are performing tremolo. Finally, if they are used to shape the timbre of a sound (e.g. by modulating the cutoff of a filter), they are performing what is sometime called wobble.

Other tools for articulation are slew limiters, which smooth step-like transitions of a control voltage. A typical use is the smoothing of a keyboard control voltage that provides a *glide* or *portamento* effect by prolonging the transition from one pitch value to another.

A somewhat related type of module is the sample and hold (S&H). This module does the inverse of a slew limiter by taking the value at given time instants and holding it for some time, giving rise to a step-like output. The operation of an S&H device is mathematically known as a zero-order hold filter. An S&H device requires an input signal and depends on a clock that sends triggering pulses. When these are received, the S&H outputs the instantaneous input signal value and holds it until a new trigger arrives. Its output is inherently step-like and can be used to control a range of other modules.

1.3.4 "Fire at Will," or in Short: Sequencers

Step sequencers is another family of modules that allow you to control the performance. Sequencers specifically had – and still have – a distinctive role in the making of electronic music, thanks to their machine-like precision and their obsessive repetition on standard time signatures. Sequencers are made of an array or a matrix of steps, each representing equally spaced time divisions. For drum machines, each step stores a binary information: fire/do not fire. The sequencer cycles repeatedly along the steps and fires whenever one of them is armed. We may call this a binary sequencer. For synthesizers, each step has one or more control voltage values associated, selectable through knobs or sliders. These can be employed to control any of the synth parameters, most notably the pitch, which is altered cyclically, following the values read at each step. Sequencers may also include both control voltage and a binary switch, the latter for arming the step. Skipping some steps allows creating pauses in the sequence.

Sequencers are usually controlled by a master clock at metronome rate (e.g. 120 bpm), and at each clock pulse a new step is selected for output, sending the value or values stored in that step. This allows, for example, storing musical phrases if the value controls the pitch of a VCO, or storing time-synchronized modulations if the value controls other timbre-related devices. Typical sequencers consist of an array of 8 or 16 steps, used in electronic dance music (EDM) genres to store a musical phrase or a drumming sequence of one or two bars with time signature 4/4. The modular market, however, provides all sorts of weird sequencers that allow for generative music, polyrhythmic composition, and so on.

Binary sequencers are used for drum machines to indicate whether a part of the drum should fire or not. Several rows are required, one for each drum part. Although the Roland TR-808 is widely recognized as one of the first drum machines that could be programmed using a step sequencer, the first drum machine ever to host a step sequencer was the Eko Computer Rhythm, produced in 1972 and developed by Italian engineers Aldo Paci, Giuseppe Censori, and Urbano Mancinelli. This sci-fi wonder has six rows of 16 lit switches, one per step. Each row can play up to two selectable drum parts (Figure 1.9).

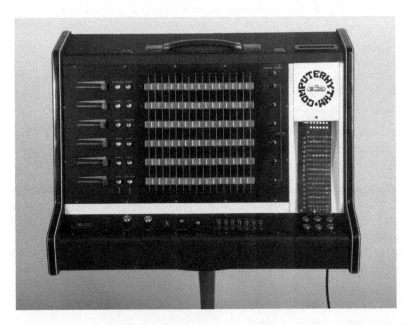

Figure 1.9: The Eko Computer Rhythm, the first drum machine ever to be programmed with a step sequencer. It was devised and engineered not more than 30 km away from where this book was written. Photo courtesy of Acusmatiq MATME. Owner: Paolo Bragaglia. Restored by Marco Molendi.

1.3.5 Utility Modules

There are, finally, a terrific number of *utility* modules that, despite their simplicity, have a high value for patching. Attenuators and attenuverters, mixers, multiples, mutes, and multiplexers and demultiplexers are very important tools to operate on signals. A brief definition is given for each one of these:

- *Attenuators and attenuverters.* An attenuator is a passive or active circuit that just attenuates the signal using a potentiometer. In the digital domain, this is equivalent to multiplying a signal by any value in the range [0, 1]. Attenuverters, additionally, are able to invert the signal, as if multiplying the signal by a number in the range [−1, 1]. Please note that inversion of a periodic signal is equivalent to a phase shift of 180° or π.
- *Mixers.* These modules allow you to sum signals together. They may be passive, providing just an electrical sum of the input voltages, they may be active, and they may have faders to control the gain of each input channel. Of course, in VCV Rack, there will be no difference between active and passive; we will just be summing discrete-time signals.
- *Multiples.* It is often useful to duplicate a signal. Multiples are made for this. They provide one input signal into several outputs. In Rack, this is not always required, since cables can be stacked from outputs, allowing duplication without requiring a multiple. However, they can still be useful to make a patch tidy.
- *Mutes.* It is sometimes useful to mute a signal, especially during a performance. Mutes are just switches that allow the signal flow from an input to an output or not.
- *Multiplexers and demultiplexers.* These modules allow for complex routing of signals. A multiplexer, or *mux*, has one input and multiple outputs and a knob to select where to route the input signal. A demultiplexer, or *demux*, on the contrary, has multiple inputs and one output. In this case, the knob selects which input to route to the output. Mux and demux devices only allow one signal to pass at a time.

Interface and control modules are also available to control a performance with external tools or add expressiveness. MIDI-to-CV modules are necessary to transform Musical Instruments Digital Interface (MIDI) messages into a CV. Theremin-like antennas and metal plates are used as input devices, while piezoelectric transducers are used to capture vibrations and touch, to be processed by other modules.

Notes

1 An early keyboard-controlled vacuum-tube synthesizer.
2 Indeed, many technological outbreaks in the musical field are derived from previous works in the communication field: from filters to modulators, from speech synthesis to digital signal processors, there is a lot we take from communication engineering.
3 Among the musicians that assisted Moog in his "modular" years (i.e. before the Model D would be released), we should mention at least Jean-Jacques Perrey and Wendy Carlos as early adopters and Herb Deutsch for assisting Moog in all the developments well before 1964.
4 In this case, the filter can be treated as an oscillator (i.e. a tone generator).

CHAPTER 2

Elements of Signal Processing for Synthesis

Modular synthesis is all about manipulating signals to craft an ephemeral but unique form of art. Most modules can be described mathematically, and in this chapter we are going to deal with a few basic aspects of signal processing that are required to develop modules. The chapter will do so in an intuitive way, to help readers that are passionate about synthesizers and computer music to get into digital signal processing without the effort of reading engineering textbooks. Some equations will be necessary and particularly useful for undergraduate or postgraduate students who require some more detail. The math uses the discrete-time notation as much as possible, since the rest of the book will deal with discrete-time signals. Furthermore, operations on discrete-time series are simpler to grasp, in that it helps avoid integral and differential equations, replacing these two operators with sums and differences. In my experience with students from non-engineering faculties, this makes some critical points better understood.

One of my aims in writing the book has been to help enthusiasts get into the world of synthesizer coding, and this chapter is necessary reading. The section regarding the frequency domain is the most tough; for the rest, there is only some high school maths. For the eager reader, the bible of digital signal processing is *Digital Signal Processing* (Oppenheim and Schafer, 2009). Engineers and scholars with experience in the field must forgive me for the overly simplified approach. For all readers, this chapter sets the notation that is used in the following chapters.

This book does not cover acoustic and psychoacoustic principles, which can be studied in specialized textbooks (e.g. see Howard and Angus, 2017), and it does not repeat the tedious account about sound, pitch, loudness, and timbre, which is as far from my view on musical signals as much as tonal theories are far from modular synthesizer composition theories.

2.1 Continuous-Time Signals

 TIP: Analog synthesizers and effects work with either voltage or current signals. As any physical quantity, these are continuous-time signals and their amplitude can take any real value – this is what we call an analog signal. Analog synthesizers do produce analog signals, and thus we need to introduce this class of signals.

A signal is defined as a function or a quantity that conveys some information related to a physical system. The information resides in the variation of that quantity in a specific domain. For instance, the voltage across a microphone capsule conveys information regarding the acoustic pressure applied to it, and henceforth of the environment surrounding it.

From a mathematical standpoint, we represent signals as functions of one or more independent variables. The independent variable is the one we indicate between braces (e.g. when we write $y = f(x)$, the independent variable is x). In other domains, such as image processing, there are usually two independent variables, the vertical and horizontal axes of the picture. However, most audio signals are represented in the time domain (i.e. in the form $f(t)$, with t being the time variable).

Table 2.1: Notable continuous-time signals of interest in sound synthesis

Name	Mathematical Description	Representation
Sine	$sin(2\pi ft)$	
Cosine	$cos(2\pi ft)$	
Decaying exponential	$e^{-\alpha t}, \alpha > 0$ Another less common form is: $a^t, 0 < a < 1$	
Sawtooth	$(t \bmod T)$	
White noise	Zero mean, aleatory signal	

Time can be defined as either continuous or discrete. Physical signals are all continuous-time signals; however, discretizing the time variable allows for efficient signal processing, as we shall see later.

Let us define an *analog signal* as a signal with continuous time and continuous amplitude:

$$s = f(t) : \mathbb{R} \to \mathbb{R}$$

The notation indicates that the variable t belongs to the real set and maps into a value that is function of t and equally belongs to the real set. The independent variable is taken from a set (in this case, $\mathbb{R}$) that is called *domain*, while the dependent variable is taken from a set that is called *codomain*. For any time instant t, $s(t)$ takes a known value.

Most physical signals, however, have finite length, and this yields true for musical signals as well, otherwise recording engineers would have to be immortal, which is one of the few qualities they still miss. For a finite continuous-time signal that lives in the interval $[T_1, T_2]$, we define it in this shorter time span:

$$s = f(t) : [T_1, T_2] \to \mathbb{R}$$

A class of useful signals is reported in Table 2.1.

2.2 Discrete-Time Signals

> **TIP**: Discrete-time signals are crucial to understand the theory behind DSP. However, they differ from digital signals, as we shall see later. They represent an intermediate step from the real world to the digital world where computation takes place.

Let us start with a question: Why do we need discrete-time signals? The short answer is that computers do not have infinite computational resources. I will elaborate on this further. You need to know that:

1. Computers crunch numbers.
2. These numbers are represented as a finite sequence of binary digits. Why finite? Think about it: Would you be able to calculate the sum of two numbers with an infinite number of digits after the decimal point? It would take you an infinite amount of time. Computers are not better than you, just a little faster. Small values are thus rounded to reduce the number of digits required to represent them. They are thus said to be finite-precision (*quantized*) numbers. Similarly, you cannot express larger numbers with a finite number of digits. This is also very intuitive: If your pocket calculator only has five digits, you cannot express a number larger than 99,999, right?
3. Computing is done in a short but not null time slice. For this reason, the less data we feed to a computer, the shorter it takes to provide the result. A continuous-time signal has infinite values between any two time instants, even very close ones. This means that it would take an infinite amount of time to process even a very small slice of signal. To make computation feasible, therefore, we need to take snapshots of the signal (*sampling*) at regular intervals. This is an

approximation of the signal, but a good one, if we set the sampling interval according to certain laws, which we shall review later.

In this chapter, we shall discuss mainly the third point (i.e. sampling). Quantization (the second point) is a secondary issue for DSP beginners, and is left for further reading. Here, I just want to point out that if you are not familiar with the term "quantization," it is exactly the same thing you do when measuring the length of your synth to buy a new shelf for it. You take a reference, the measuring tape, and compare the length of the synth to it. Then you approximate the measure to a finite number of digits (e.g. up to the millimeter). Knowing the length with a precision up to the nanometer is not only unpractical by eye, but is also useless and hard to write, store, and communicate. Quantization has only marginal interest in this book, but a few hints on numerical precision are given in Section 2.13.

Let us discuss sampling. As we have hinted, the result of the sampling process is a discrete-time signal (i.e. a signal that exists only at specific time instants). Let us now familiarize ourselves with discrete-time signals. These signals are functions, like continuous-time signals, with the independent variable not belonging to the real set. It may, for example, belong to the integer set $\mathbb{Z}$ (i.e. to the set of all positive and negative integer numbers). While real signals are defined for any time instant t, discrete signals are defined only at equally spaced time instants n belonging to $\mathbb{Z}$. This set is less populated than $\mathbb{R}$ because it misses values between integer values. There are, for instance, infinite values between instants $n = 1$ and $n = 2$ in the real set that the integer set does not have. However, do not forget that even $\mathbb{Z}$ is an infinite set, meaning that theoretically our signal can go on forever.

Getting more formal, we define a discrete-time signal as $s = f[n] : \mathbb{Z} \to \mathbb{R}$, where we adopt a notation usual for DSP books, where discrete-time signals are denoted by the square brackets and the use of the variable n instead of t. As you may notice, the signal is still a real-valued signal. As we have discussed previously, another required step is quantization. Signals with their amplitude quantized do not belong to $\mathbb{R}$ anymore. Amplitude quantization is a step that is independent of time discretization. Indeed, we could have continuous time signals with quantized amplitude (although it is not very usual). Most of the inherent beauty and convenience of digital signal processing is related to the properties introduced by time discretization, while amplitude quantization has only a few side effects that require some attention during the implementation phase. We should state clearly that digital signals are both discretized in time and their amplitude. However, for simplicity, we shall now focus on signals that have continuous amplitude.

A discrete-time signal is a sequence of ordered numbers and is generally represented as shown in Figure 2.1. Any such signal can be decomposed into single pulses, known as Dirac pulses. A Dirac pulse sequence with unitary amplitude is shown in Figure 2.2, and is defined as:

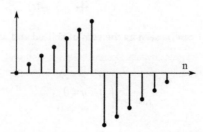

Figure 2.1: An arbitrary discrete-time signal.

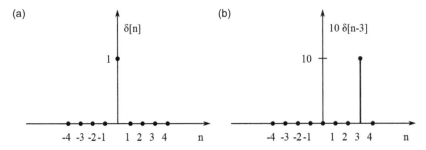

Figure 2.2: The Dirac sequence $\delta[n]$ (a) and the sequence $10\delta[n-3]$ (b).

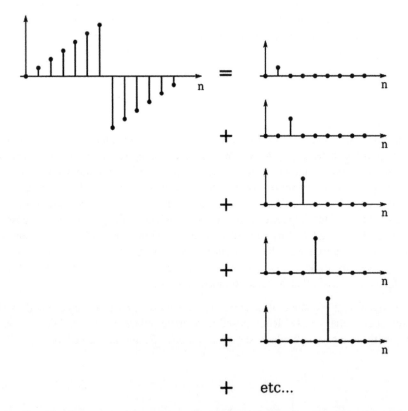

Figure 2.3: The signal in Figure 2.1 can be seen as the sum of shifted and weighted copies of the Dirac delta.

$$\delta[n] = \begin{cases} 0 & n \neq 0 \\ 1 & n = 0 \end{cases} \qquad (2.1)$$

A Dirac pulse of amplitude A shifted in time by T samples is denoted as:[1]

$$A\delta[n - T] = \begin{cases} 0 & n \neq T \\ A & n = T \end{cases} \tag{2.2}$$

By shifting in time multiple Dirac pulses and properly *weighting* each one of them (i.e. multiplying by a different real coefficient), we obtain any arbitrary signal, such as the one in Figure 2.1. This process is depicted in Figure 2.3. The Dirac pulse is thus a sort of elementary particle in this quantum game we call DSP.[2] If you have enough patience, you can sum infinite pulses and obtain an infinite-length discrete-time signal. If you are lazy, you can stop after N samples and obtain a finite-length discrete-time signal $s = f[n] : [0, N] \rightarrow \mathbb{R}$. If you are lucky, you can give a mathematical definition of a signal and let it build up indefinitely for you. This is the case of a sinusoidal signal, which is infinite-length, but it can be described by a finite set of samples (i.e. one period).

Table 2.2 reports notable discrete-time sequences that are going to be of help in progressing through the book.

Table 2.2: Notable discrete-time sequences, their mathematical notation and their graphical representation

	Mathematical description	Graphical representation
Unitary step	$u[n] = \begin{cases} 0 & n < 0 \\ 1 & n \geq 0 \end{cases}$	
Discrete-time ramp	$r[n] = n$	
Discrete-time sine	$sin(2\pi f n)$	
Discrete-time decaying exponential	$e^{-\alpha n}, \alpha > 0$	

2.3 Discrete-Time Systems

TIP: Discrete-time systems must not be confused with digital systems. As with discrete-time signals, they represent an intermediate step between continuous-time systems and digital systems. They have a conceptual importance, but we do not find many of them in reality. One notable exception is the Bucket-Brigade delay. This kind of delay, used in analog effects for chorus, flangers, and other modulation effects, samples the continuous-time input signal at a frequency imposed by an external clock. This will serve as a notable example in this section.

How do we deal with signals? How do we process them? With systems!

Systems are all around us. A room is an acoustic system that spreads the voice of a speaker. A loudspeaker is an electroacoustic system that transforms an electrical signal into an acoustic pressure wave. A *fuzz* guitar pedal is a nonlinear system that outputs a distorted version of an input signal. A guitar string is a damped system that responds to a strike by oscillating with its own modes. The list can go on forever, but we need to focus on those systems that are of interest to us. First of all, we are interested in discrete-time systems, while the above lists only continuous-time systems. Discrete-time systems can emulate continuous-time systems or can implement something new. Discrete-time systems are usually implemented as algorithms running on a processing unit. They are depicted in the form of unidirectional signal-flow graphs, such as the one seen in Figure 2.4, and they are usually implemented in code.

Discrete-time systems can:

- have one or more inputs and one or more outputs;
- be stable or unstable;
- be linear or nonlinear;
- have memory or be static; and
- be time-variant or time-invariant.

Many more discrete-time systems exist, but we shall focus on the above ones. In digital signal processing textbooks, you usually find a lot on *linear time-invariant systems* (LTIs). These are

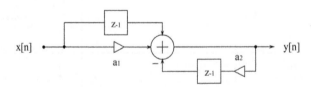

Figure 2.4: Unidirectional graph for an LTI system, composed by a summation point, unitary delays (z^{-1}), and products by scalar (gains), depicted with a triangle. The graph implements the difference equation
$$y[n] = a_1x[n] + x[n-1] - a_2y[n-1].$$

fundamental to understand the theory underlying systems theory, control theory, acoustics, physical modeling, circuit theory, and so on. Unfortunately, useful systems are nonlinear and time-variant and often have many inputs and/or outputs. To describe such systems, most theoretical approaches try to look at them in the light of the LTI theory and address their deviations with different approaches. For this reason, we shall follow the textbook approach and prepare some background regarding the LTI systems. Nonlinear systems are much more complex to understand and model. These include many acoustical systems (e.g. hammer-string interaction, a clarinet reed, magnetic response to displacement in pickups and loudspeaker coils) and most electronic circuits (amplifiers, waveshapers, etc.). However, to deal with these, a book is not enough, let alone a single chapter! A few words on nonlinear systems will be spent in Sections 2.11 and 8.3. Further resources will be given to continue on your reading and start discrete-time modeling of nonlinear systems.

Now it is time to get into the mathematical description of LTI systems and their properties. A single-input, single-output discrete-time system is an operator $T\{\cdot\}$ that transforms an input signal $x[n]$ into an output signal $y[n]$:

$$y[n] = T\{x[n]\} \tag{2.3}$$

Such a system is depicted in a signal-flow graph, as shown in Figure 2.5.
There are three basic building blocks for LTI systems:

- scalar multiplication;
- summation points; and
- delay elements.

The delay operation requires an explanation. Consider the following equation:

$$y[n] = x[n - L] \tag{2.4}$$

with the time index n running possibly from minus infinity to infinity and L being an integer. Equation 2.4 describes a delaying system introducing a delay of length L samples. You can prove yourself that this system delays a signal with pen and paper, fixing the length L and showing that for each input value the output is shifted by L samples to the right. The delay will be often denoted as z^{-L}, for reasons that are clear to the expert reader but are less apparent to other readers. This is related to the Z-transform (the discrete-time equivalent of the Laplace transform), which is not covered in this book for simplicity. Indeed, the Z-transform is an important tool for the analysis and design of LTI systems, but for this hands-on book it is not strictly necessary, and as usual I recommend reading a specialized DSP textbook.

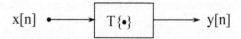

Figure 2.5: Flow graph representation of a discrete-time system.

Let us now take a break and discuss the Bucket-Brigade delay (BBD). This kind of device consists of a cascade of L one-sample delay cells (z^{-1}), creating a so-called delay line (z^{-L}). Each of these cells is a capacitor, able to store a charge, and each cell is separated by means of transistors acting as a gate. These gates are activated at once by a periodic clock, thus allowing the charge of each cell to be transferred, or *poured*, to the next one. In this way, the charge stored in the first cell travels through all the cells, reaching the last one after L clock cycles, thus delayed by L clock cycles. You can clearly see the analogy with a bucket of water that is passed from hand to hand in a line of firefighters having to extinguish a fire. Figure 2.6 may help you understand this concept. This kind of device acts as a discrete-time system; however, the stored values are physical quantities, and thus no digital processing happens. However, being a discrete-time system, it is subject to the sampling theorem (discussed later) and its implications.

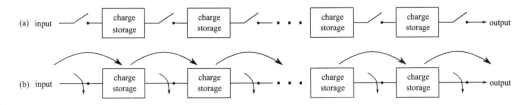

Figure 2.6: Simplified diagram of a Bucket-Brigade delay. Charge storage cells are separated by switches, acting as gates that periodically open, letting the signals flow from left to right. When the switches are open (a), no signal is flowing, letting some time pass. When the switches close (b), each cell transmits its value to the next one. At the same time, the input signal is sampled, or *freezed*, and its current value is stored in the first cell. It then travels the whole delay line, through the cells, one by one.

Back to the theory, we have now discussed three basic operations: sum, product, and delay. By composing these three operations, one can obtain all sorts of LTI systems, characterized by the linearity and the time-invariance properties. Let us examine these two properties.

Linearity is related to the principle of superposition. Let $y_1[n]$ be the response of a system to input $x_1[n]$, and similarly $y_2[n]$ be the response to input stimulus $x_2[n]$. A system is linear if – and only if – Equations 2.5 and 2.6 hold true:

$$T\{x_1[n] + x_2[n]\} = T\{x_1[n]\} + T\{x_2[n]\}$$
$$= y_1[n] + y_2[n] \quad \text{(additivity property)} \tag{2.5}$$

$$T\{ax[n]\} = aT\{x[n]\} = ay[n] \quad \text{(scaling property)} \tag{2.6}$$

where a is a constant. These properties can be combined into the following equation, stating the principle of superposition:

$$T\{ax_1[n] + bx_2[n]\} = T\{ax_1[n]\} + T\{bx_2[n]\} \tag{2.7}$$

for arbitrary constants a and b.

In a few words, a system is linear whenever the output to each different stimulus is independent on other additive stimuli and their amplitudes.

LTI systems are only composed of sums, scalar products, and delays. Any other operation would violate the linearity property. Take, for instance, a simple system that instead of a scalar product multiplies two signals. This is described by the following difference equation:

$$y[n] = x[n] * x[n] = x[n]^2 \tag{2.8}$$

This simple system is memoryless (the output is not related to a previous input) and nonlinear as the exponentiation is not a linear function. It is straightforward that $y[n] = (x_1[n] + x_2[n])^2 \neq x_1[n]^2 + x_2[n]^2$, except for trivial cases with at least one of the two inputs equal to zero.

Another interesting property of LTI systems is *time-invariance*. A system is time-invariant if it always operates in the same way disregarding the moment when a stimulus is provided, or, more rigorously, if:

$$y[n] = T\{x[n]\} \text{ then } y[n - n_0] = T\{x[n - n_0]\}, \quad \forall n_0, x[n] \tag{2.9}$$

Time-variant systems do change their behavior depending on the moment we are looking at them.

Think of an analog synthesizer filter affected by changes in temperature and humidity of the room, and thus behaving differently depending on the time of the day. We can safely say that such a system is a (slowly) time-variant one. But even more important, it can be operated by a human and changes its response while the user rotates its knobs. It is thus (quickly) time-variant while the user modifies its parameters. A synthesizer filter, therefore, is only an approximation of an LTI system: it is time-variant. We can neglect slow variations due to the environment (digital filters are not affected by this – that's one of their nice properties), and assume time-invariance while they are not touched, but we cannot neglect the effect of human intervention. This, in turn, calls for a check on stability while the filter is manipulated.

Causality is another property, stating that the output of a system depends only on any of the current and previous inputs. An anti-causal system can be treated in mathematical terms but has no real-time implementation in the real world since nobody knows the future (yet).

Anti-causal systems can be implemented offline (i.e. when the entire signal is known). A simple example is the reverse playback of a track. You can invert the time axis and play the track from the end to the beginning because it has been entirely recorded. Another example is a reverse reverb effect. Real-time implementations of this effect rely on storing a short portion of the input signal, reverting it, and then passing through a reverb. This is offline processing as well, because the signal is first stored and then reverted.

Stability is another important property of systems. A so-called bounded-input, bounded-output (BIBO) stable system is a system whose output never goes to infinity when finite signals are applied:

$$\text{if } |x[n]| \leq B_x < +\infty \text{ then } |y[n]| \leq B_y < +\infty \tag{2.10}$$

This means that if the input signal is bounded (both in the positive and negative ranges) by a value B_x for its whole length, then the output of the system will be equally bounded by another maximum value B_y, and both the bounding values are not infinite. In other words, the output will never go to infinity, no matter what the input is, excluding those cases where the input is infinite (in such cases, we can accept the output to go to infinity).

Unstable systems are really bad for your ears. An example of an unstable system is the feedback connection of a microphone, an amplifier, and a loudspeaker. Under certain conditions (the Barkhausen stability criterion), the system becomes unstable, growing its output with time and creating a so-called Larsen effect. Fortunately, no Larsen grows up to infinity, because no public address system is able to generate infinite sound pressure levels. It can hurt, though. In Gabrielli et al. (2014), a virtual acoustic feedback algorithm to emulate guitar howling without hurting your ears is proposed and some more information on the topic is given.

One reason why we are focusing our attention on LTI systems is the fact that they are quite common and they are tractable with a relatively simple math. Furthermore, they are uniquely defined by a property called *impulse response*. The impulse response is the output of the system after a Dirac pulse is applied to its input. In other words, it is the sequence:

$$h[n] = T\{\delta[n]\} \tag{2.11}$$

An impulse response may be of finite or infinite length. This distinction is very important and leads to two classes of LTI systems: infinite-impulse response system (IIR) and finite-impulse response system (FIR). Both are very important in the music signal processing field.

Knowing the impulse response of a system allows to compute the output as:

$$y[n] = T\{x[n]\} = \sum_{k=-\infty}^{+\infty} h[k] \cdot x[n-k] \tag{2.12}$$

Or, if the impulse response has finite length M:

$$y[n] = T\{x[n]\} = \sum_{k=0}^{M} h[k] \cdot x[n-k] \tag{2.13}$$

Those who are familiar with math and signal processing will recognize that the sum in Equation 2.12 is the convolution between the input and the impulse response:

$$y[n] = x[n] * h[n] \tag{2.14}$$

where the convolution is denoted by the symbol *. As you may notice, the convolution sum can be quite expensive if the impulse response is long: for each output sample, you have to evaluate up to infinite products! Fortunately, there are ways to deal with IIR systems that do not require infinite operations per sample (indeed, most musical filters are IIR, but very cheap). Inversely, FIR systems with a very long impulse response are hard to deal with (partitioning the convolution can help by exploit parallel processing units). In this case, dealing with signals in the frequency domain provides some help, as the convolution between $x[n] * h[n]$ can be seen in the frequency domain as the product of their Fourier transforms (this will be shown in Section 2.4.3). Of particular use are those LTI systems where impulse response can be formulated in terms of differential equations of the type:

$$\sum_{k=0}^{N} a_k y[n-k] = \sum_{k=0}^{M} b_k x[n-k]. \tag{2.15}$$

These LTI systems, most notably filters, are easy to implement and analyze. As you can see, they are based on three operations only: sums, products, and delays. The presence of delay is not obvious from Equation 2.15, but consider that to recall the past elements $(x[n-1], x[n-2]$, etc.), some sort of delay mechanism is required. This requires that the present input is stored and recalled after k samples. Delays are obtained by storing the values in memory. Summing, multiplying, writing a value to memory, reading a value from memory: these operations are implemented in all modern processors[3] by means of dedicated *instructions*. Factoring the LTI impulse response as in Equation 2.15 helps making IIR filters cheap: there is no need to perform an infinite convolution: storing the previous outputs allows you to store the memory of all the past history.

2.4 The Frequency Domain

Before going deeper and explaining how to transform a continuous-time signal into a discrete-time signal, and vice versa, we need to introduce a new domain: the frequency domain.

The physical concept of frequency itself is very simple. The frequency is the number of occurrences per unit of time of a given phenomenon (e.g. the repetition of a cycle of a sine wave or the zero crossing of a generic signal). This is generally called temporal frequency and denoted with the letter f. The measurement unit for the temporal frequency is the Hertz, Hz (with capital H, please!), in memory of Heinrich Hertz. The reciprocal of the frequency is the period T (i.e. the time required to complete one cycle at the given frequency), so that $T = 1/f$.

While audio engineers are used to describing the frequency in Hz, because they deal with acoustic signals, in signal processing we are more used to *angular frequency*, denoted with the Greek letter ω and measured in *radians* per second (rad/s).[4] For those more used to angular degrees, 1 rad $\simeq 57.2°$ and 2π rad $= 360°$. The angular frequency is related to the temporal frequency, but it measures the angular displacement per unit of time. If we take, for example, a running wheel, the angular frequency measures the rotation per second, while the temporal frequency measures the number of complete rotations per second. In other words if the wheel completes a turn in a second, $f = 1$ Hz, $\omega = 2\pi$. The analytical relation between temporal frequency, angular frequency, and period is thus:

$$f = \frac{\omega}{2\pi}, \quad \text{i.e. } \omega = 2\pi f = \frac{2\pi}{T}. \tag{2.16}$$

If the concept of frequency is clear, we can now apply this to signal processing. Follow me with a little patience as I point out some concepts and then connect them together. In signal processing, it is very common to use *transforms*. These are mathematical operators that allow you to observe the signal under a different light (i.e. they change shape[5] to the input signal and take it to a new domain, without affecting its informative content). The domain is the realm where the signal lives. Without going too abstract, let us take the mathematical expression:

$$s(x) = \sin(\omega x) \tag{2.17}$$

If $x \in \mathbb{R}$, then the resulting signal $s(x)$ lives in the continuous-time domain. If $x \in \mathbb{Z}$, then $s(x)$ lives in the discrete-time domain. We already know of these two domains. We are soon going to define a third domain, the frequency domain. We shall see that in this new domain, the sine signal of Equation 2.17 has a completely new shape (a line!) but still retains the same meaning: a sinusoidal wave oscillating at ω rad/s.

There is a well-known meme from the Internet showing that a solid cylinder may look like a circle or a square if projected on a wall with light coming from a side or from the bottom, stating that observing one phenomenon from different perspectives can yield different results, yet the truth is a complex mixture of both observations (see Figure 2.7). Similarly, in our field, the same phenomenon (a signal) can be observed and projected in different domains. In each domain, the signal looks different, because each domain describes the signal in a different way, yet both domains are speaking of the same signal. In the present case, one domain is time and the other is frequency. Often the signal is generated in the

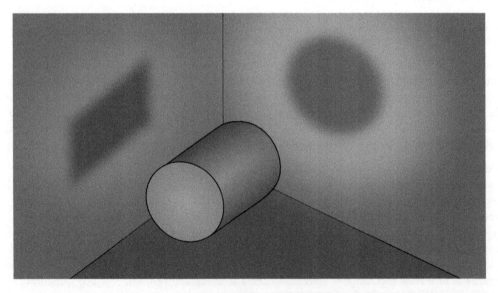

Figure 2.7: The projection of a cylinder on two walls from orthogonal perspectives. The object appears with different shapes depending on the point of view. Similarly, any signal can be projected in the time and the frequency domain, obtaining different representations of the same entity.

time domain, but signals can be synthesized in the frequency domain as well. There are also techniques to observe the signal in both domains, projecting the signal in one of the known time-frequency domains (more in Section 2.4.6), similar to the 3D image of the cylinder in Figure 2.7. Mixed time-frequency representations are often neglected by sound engineers, but are very useful to grasp the properties of a signal at a glance, similar to what the human ear does.[6]

We shall first discuss the frequency domain and how to take a time-domain signal into the frequency domain.[7]

2.4.1 Discrete Fourier Series

Let us consider a discrete-time signal. We have just discovered that we can see any discrete-time signal as a sum of Dirac pulses, each one with its weight and time shift. The Dirac pulse is our elementary brick to build a discrete-time signal. It is very intuitive to see how from the sum of these bricks (Figure 2.3) we build the signal in Figure 2.1. Let us now consider a different perspective. What if we consider the signal in Figure 2.1 as composed by a sum of sinusoidal signals? Figure 2.8 shows how this happens. It can be shown that this process is general, and a large class of signals (under the appropriate conditions) can be seen as consisting of elementary sinusoidal signals, each with its weight, phase, and frequency. This analysis is done through different formulations of the so-called Fourier transform, depending on the kind of input signal we consider. We are going to start with a periodic discrete-time signal.

Let us take a periodic[8] signal $\tilde{x}[n]$. We shall take an ideal sawtooth signal, for the joy of our synthesizer-fanatic reader. This signal is created by repeating linear ramps, increasing the dependent variable over time, and resetting it to zero after an integer number of samples N. This signal takes the same value after an integer number of samples, and can thus be defined as *periodic*:

$$\tilde{x}[n] = \tilde{x}[n+N] = \tilde{x}[n+2N] = \cdots = \tilde{x}[n+kN] \; \forall \; k \in \mathbb{N} \tag{2.18}$$

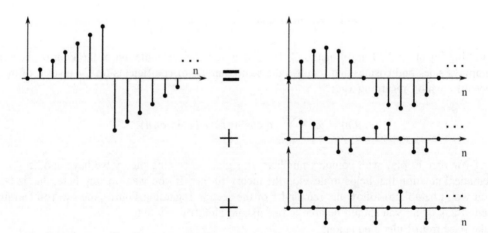

Figure 2.8: The signal of Figure 2.1 is periodic, and can thus be decomposed into a sum of periodic signals (sines).

As we know, the frequency of this signal is the reciprocal of the period (i.e. $f_F = 1/N$, i.e. $\omega_F = 2\pi/N$). In this section, we will refer to the frequency as the angular frequency. The fundamental frequency is usually denoted with a 0 subscript, but for a cleaner notation we are sticking with the F subscript for the following lines.

Given a periodic signal with period N, there are a small number of sinusoidal signals that have a period equal or integer divisor of N. What if we write the periodic signal as a sum of those signals? Let us take, for example, cosines as these basic components. Each cosine has angular frequency ω_i that is an integer multiple of the fundamental angular frequency ω_F. Each cosine will have its own amplitude and phase. The presence of a constant term a_0 is required to take into account an offset, or *bias*, or DC component (i.e. a component at null frequency – it is not oscillating, but it fulfills Equation 2.18). This is the *real-form discrete Fourier series (DFS)*.

$$\tilde{x}[n] = a_0 + \sum_{k=1}^{N-1} a_k \cos(\omega_k n + \phi_k), \omega_k = k\omega_F \tag{2.19}$$

Equation 2.19 tells us that *any periodic signal* in the discrete-time domain can be seen as a sum of a finite number of cosines at frequencies that are multiples of the fundamental frequency, each weighted properly by an amplitude coefficient a_k and time-shifted by a phase coefficient ϕ_k. The cosine components are the harmonic partials,[9] or *harmonics*, of the original signal, and they represent the *spectrum* of the signal.

We can further develop our math to introduce a more convenient notation. Bear with me and Mr. Jean Baptiste Joseph Fourier for a while. It can be shown that a cosine having frequency and phase components in its argument can be seen as the sum of a sine and a cosine of the same frequency and with no phase term in their argument, as shown in Equation 2.20. In other words, the added sinusoidal term takes into account the phase shift given by ϕ. After all, intuitively, you already know that cosine and sine are the same thing, just shifted by a phase offset:

$$c \cos(\omega + \phi) = a \sin(\omega) + b \sin(\omega)$$
$$\text{with } c = sgn(a)\sqrt{a^2 + b^2}, \phi = \tan^{-1}\left(\frac{-b}{a}\right) \tag{2.20}$$

We can plug Equation 2.20 into Equation 2.19, obtaining a new formulation of the Fourier theorem that employs sines and cosines and requires the same number of coefficients (a_k, b_k instead of a_k, ϕ_k) to store information regarding $\tilde{x}[n]$:

$$\tilde{x}[n] = a_0 + \sum_{k=1}^{N} [a_k \cos(\omega_k n) + b_k \sin(\omega_k n)] \tag{2.21}$$

Now, if you dare to play with complex numbers, in signal processing theory we have another mathematical notation that helps to develop the theory further. If you want to stop here, this is OK, but you won't be able to follow the remainder of the section regarding Fourier theory. You can jump to Section 2.4.3, and you should be able to understand almost everything.
We take pace from Euler's equation:

$$e^{j\theta} = \cos\theta + j\sin\theta \tag{2.22}$$

where j is the imaginary part.[10] The term $e^{j\theta}$ is called a complex exponential.

By reworking Equation 2.22, we obtain the following equalities:

$$
\begin{aligned}
\cos\theta &= \frac{e^{j\theta} + e^{-j\theta}}{2} = \frac{1}{2}e^{j\theta} + \frac{1}{2}e^{-j\theta} \\
\sin\theta &= \frac{e^{j\theta} - e^{-j\theta}}{2j} = \frac{-1}{2}je^{j\theta} + \frac{1}{2}je^{-j\theta}
\end{aligned}
\tag{2.23}
$$

Now we shall apply these expressions to the second formulation of the DFS. If we take $\theta = \omega_k n$, we can substitute the sine and cosine expressions of Equation 2.23 in Equation 2.21, yielding:

$$\tilde{x}[n] = a_0 + \sum_{n=1}^{N}\left(a_n\left(\frac{1}{2}e^{j\theta} + \frac{1}{2}e^{-j\theta}\right) + b_n\left(-\frac{1}{2}je^{j\theta} + \frac{1}{2}je^{-j\theta}\right)\right)$$

This can be rewritten by separating the two sums as:

$$\tilde{x}[n] = a_0 + \sum_{n=1}^{N}\left(\left(\frac{1}{2}a_n - \frac{1}{2}jb_n\right)e^{j\theta}\right) + \sum_{n=1}^{N}\left(\left(\frac{1}{2}a_n + \frac{1}{2}jb_n\right)e^{-j\theta}\right)$$

The last step consists of transforming all the constant terms a_n, b_n into one single vector that shall be our frequency representation of the signal:

$$\tilde{x}[n] = \sum_{k=-N}^{N}\tilde{X}[k]e^{j\theta} = \sum_{k=-N}^{N}\tilde{X}[k]e^{j\omega_k n} \tag{2.24}$$

where

$$\tilde{X}[k] = \begin{cases} \frac{1}{2}a_k - \frac{1}{2}jb_k, & k \geq 1 \\ \frac{1}{2}a_0, & k = 0 \\ \frac{1}{2}a_{|k|} + \frac{1}{2}jb_{|k|}, & k \leq -1 \end{cases} \tag{2.25}$$

This little trick not only makes Equation 2.21 more compact, but also instructs us how to construct a new discrete signal $\tilde{X}[k]$ that bears all the information related to $\tilde{x}[n]$.

Several key concepts can be drawn by observing how $\tilde{X}[k]$ is constructed:

- $\tilde{X}[k]$ is complex valued except from $\tilde{X}[0]$.
- $\tilde{X}[0]$ is the offset component, now seen as a null-frequency component, or DC term.
- $\tilde{X}[k]$ is composed of terms related to negative frequency components ($k \leq -1$) and positive frequency components ($k \geq 1$). Negative frequencies have no physical interpretation, but nonetheless are very important in DSP.
- Negative frequency components have identical coefficients, besides the sign of b_k (i.e. they are complex conjugate $\tilde{X}[k] = \tilde{X}[-k]^*$). Roughly speaking, negative frequency components can be neglected when we observe a signal.[11]

To conclude, we are now able to construct a frequency representation of a time-domain signal for periodic signals. The process is depicted in Figure 2.9.

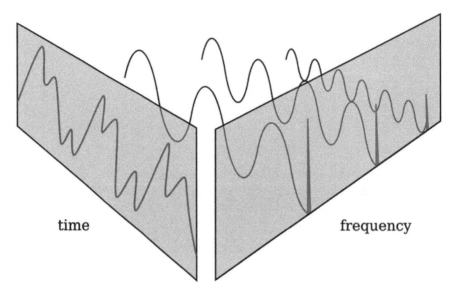

Figure 2.9: A signal in time (left), decomposed as a sum of sinusoidal signals (waves in the back). The frequency and amplitudes of the sinusoidal signals are projected into a frequency domain signal (right). The projection metaphor of Figure 2.7 is now put in context.

2.4.2 Discrete Fourier Transform

Theoretically speaking, periodic signals are a very important class of signals in music. However, periodic signals stay exactly the same for their whole life (i.e. from minus infinity to infinity). The closest to such a signal I can think of is a Hammond tonewheel that spins forever. But eventually AC mains will blackout for a while – at least at my institution it happens quite frequently. A broader class of signals of interest are non-periodic and/or non-stationary (i.e. changing over time). To describe non-periodic but stationary signals, we need to depart from the DFS and develop the discrete Fourier transform.[12]

We shall here extend the concepts developed for the DFS to non-periodic signals of finite duration (i.e. those signals that are non-zero only for a finite period of time). Let us take a signal $x[n]$, which is null besides a finite time interval n (e.g. a short ramp), as in Figure 2.10a.

By applying a neat trick, our efforts for obtaining the DFS will not be wasted. Indeed, if we make $x[n]$ periodic, by replicating it infinite times we can apply the DFS to the periodic version of the signal $\tilde{x}[n]$, which we highlight with a tilde, as before, to highlight that it has been made periodic. The periodic signal can be treated using the DFS as shown before. The aperiodic signal is again seen as a sum of complex exponentials (Equation 2.26). The number of the exponentials is equal to the finite length of the signal. This time, the complex exponentials are called *partials* and are not necessarily harmonic. The set of the complex-valued coefficients $X[k]$ is again called a *spectrum*. The spectrum is a discrete signal with values (often called *bins*) defined only for the finite set of *frequencies k*:

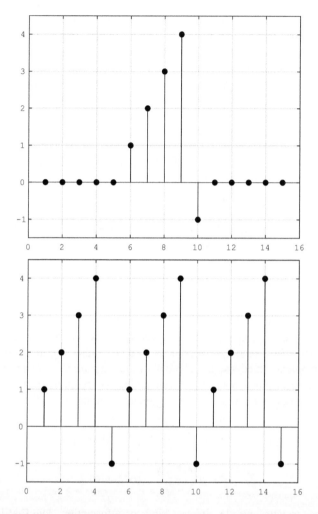

Figure 2.10: A finite duration signal $x[n]$ (a) is converted into a periodic signal $\tilde{x}[n]$ by juxtaposing replicas (b). The periodic version of $x[n]$ can be now analyzed with the DFS.

$$x[n] = \sum_{k=0}^{N-1} X[k] e^{-j(2\pi/N)kn} \tag{2.26}$$

Please note that in Equation 2.26, we do not refer to a fundamental frequency ω_k, since $x[n]$ is not periodic, but we made the argument of the exponential explicit, where $k(2\pi/N)$ are the $(N-1)$ frequencies of the complex exponentials and N can be selected arbitrarily, as we shall later detail.

Equation 2.26 not only states that a discrete-time signal can be seen as a sum of elementary components. It also tells how to *transform* a spectrum $X[k]$ into a time-domain signal. But wait a minute: we do not yet know how to obtain the spectrum! For tutorial reasons, we focused on understanding the Fourier series and its derivatives. However, we need, in practice, to know how to transform a signal from the time domain to the frequency domain. The *discrete Fourier transform* (DFT) is a mathematical operator that takes a real[13] discrete-time signal of length N into a complex discrete signal of length N that is function of frequency:

$$F : \mathbb{R}^N \to \mathbb{C}^N \tag{2.27}$$

The DFT is invertible, meaning that the signal in the frequency domain can be transformed back to obtain the original signal in the time domain without degradation of the signal (i.e. $F^{-1}\{F\{x[n]\}\} = x[n]$). The *direct* and *inverse* DFT are done through Equations 2.28 and 2.29:

$$X[k] = \sum_{k=0}^{N-1} x[n]e^{-j(2\pi/N)kn} \quad \text{(DFT)} \tag{2.28}$$

$$x[n] = \frac{1}{N} \sum_{k=0}^{N-1} X[k]e^{-j(2\pi/N)kn} \quad \text{(IDFT)} \tag{2.29}$$

The equations are very similar. The added term $1/N$ is required to preserve the energy. As seen above, $X[k]$ is a complex-valued signal. How to interpret this? The real and imaginary part of each frequency bin do not alone tell us anything of practical use. To gather useful information, we need to process further the complex-valued spectrum to obtain the *magnitude spectrum* and the *phase spectrum*. The magnitude spectrum is obtained by calculating the modulus (or norm) of each complex value, while the phase is obtained by the argument function:

$$\|X_k\| = \sqrt{\Re\{X_k\}^2 + \Im\{X_k\}^2} \quad \text{(Modulus)} \tag{2.30}$$

$$\angle X_k = \arctan\left(\frac{\Im X_k}{\Re X_k}\right) \quad \text{(Argument)} \tag{2.31}$$

The magnitude and phase spectra provide different kinds of information. The first is most commonly used in audio engineering because it provides information that we directly understand (i.e. the energy of the signal at each frequency component, something our ears are very sensitive to). The phase spectrum, on the other hand, tells the phase of the complex exponentials along the frequency domain. It has great importance in the analysis of LTI systems.
It is important to note that the length of the spectrum is equal to the length of the signal in the time domain.

Since the number of DFT points, or DFT *bins* (i.e. the number of frequencies), is equal to the length N of the signal, we incur a practical issue: very short signals have very few DFT bins (i.e. our knowledge of the signal in frequency is quite rough). To gather more insight, there is a trick called *zero-padding*. It consists of adding zeros at the end or at the beginning of the signal. This artificially increases the signal length to $M > N$ points by padding with zeros, thus increasing the number of DFT bins. Zero-padding is also employed to round the length of the signal to the closest power of 2 that is larger than the signal, which makes it suitable to efficient implementations of the DFT (a few words regarding computational complexity will come later). Applying zero-padding (i.e. calculating the M-points DFT) does not alter the shape of the spectrum; it just adds M-N points by interpolating the values of the N-points DFT.

So far, we have discovered:

- Any finite-length discrete-time signal can be described in terms of weighted complex exponentials (i.e. sum of sinusoids). The weights are called the spectrum of the signal. The complex exponentials are equally spaced in frequency.
- From the weights of these exponentials, we can obtain the signal back in the time domain.
- Both $x[n]$ and $X[k]$ have same length N.

Let us now review the DFT of notable signals.

2.4.3 Properties of the Discrete Fourier Transform

Linearity:

$$ax_1[n] + bx_2[n] \Leftrightarrow aX_1(k) + bX_2(k) \tag{2.32}$$

meaning that the DFT of the sum of two signals is equal to the sum of the two DFTs. This property has a lot of practical and theoretical consequences.
Time scaling:

$$F\{x(at)\} = \frac{1}{a} X\left(\frac{k}{a}\right) \tag{2.33}$$

One of the consequences of this property is of interest to sound designers: it implies the pitch shifting of a signal that is played back faster ($a > 1$) or slower ($a < 1$).
Periodicity:

$$W_N^{k(N+n)} = W_N^{(k+N)n} = \left(W_N^{kn}\right) \tag{2.34}$$

The spectrum is replicated in frequency (this is relevant to the problem of aliasing).
Convolution:

$$x[n] * y[n] \Leftrightarrow X(k) \cdot Y(k) \tag{2.35}$$

This means that the product of two DFTs is equal to the convolution between the two signals. On par with that, it is also true that the product of two signals in the time domain is equivalent to the convolution of their spectra in the frequency domain (i.e. $x[n] \cdot y[n] \Leftrightarrow X(k) * Y(k)$).
Parseval's theorem:

$$\sum_{n=-\infty}^{+\infty} |x[n]|^2 = \frac{1}{2\pi} \int_{-\pi}^{\pi} |X(e^{j\omega})|^2 d\omega$$

meaning that the energy stays the same whether we evaluate it in the time domain (left) or in the frequency domain (right)

2.4.4 Again on LTI Systems: The Frequency Response

By now, we know that a signal $x[n]$ is completely described by its DFT $X[k]$. The impulse response of a system is a signal, a very important one, because it contains all the information related to its system. But what happens if we transform it into the frequency domain? We obtain another very important property of the system, which conveys all the information regarding the system, just translated into the frequency domain. This is the so-called *frequency response*:

$$H[k] = F\{h[n]\} \tag{2.36}$$

The frequency response is even more useful than the impulse response for musical applications. If you want to design a filter, for example, you can design its desired spectrum and transform it into the time domain through the inverse DFT.[14] The frequency response of two systems can be compared to determine which one best suits your frequency specifications.

But there is one more use for the frequency response: you can use it to process a signal. Processing a signal with an LTI system implies the convolution between the signal and the system impulse response. This can be expensive in some cases. Computational savings can be obtained by transforming both in the frequency domain, multiplying them, and transforming them back in the domain. This is possible thanks to the convolution property, reported in Section 2.4.3. Processing the signal through the LTI system in such a way may reduce the computational cost, provided that our latency constraints allow us to buffer the input samples for the DFT. The larger this buffer, the larger the computational savings. Why so? Keep on reading through the next section.

Studio engineers and music technology practitioners usually confuse the terms "frequency response" and "magnitude frequency response." The frequency response obtained by computing the DFT of the impulse response is a complex-valued signal, not very informative for us humans. Computing its magnitude provides the sort of curves we are more used to seeing for evaluating loudspeaker quality or the spectral balance of a mix. However, they miss the phase information, which can be computed as well as the argument function of the frequency response.

2.4.5 Computational Complexity and the Fast Fourier Transform

The computational complexity of a DSP algorithm is the number of operations (generally sums and products) that are required for each input sample. It does not necessarily tell how fast the algorithm will run on your computer. There are many complex factors related to the implementation of an algorithm, such as the available computational resources (memory, cache, registers, mathematical instructions, parallel processing units, etc.) or the coding strategy, that are important to the fast execution of a piece of code, but in general we first evaluate the computational cost of an algorithm and try to reduce it on paper, and then see how it performs in the real world.

For the DFT, consider the following equation:

$$X[k] = \sum_{n=0}^{N-1} x[n] W_N^{kn} \quad k = 0, 1, \ldots, N-1 \tag{2.37}$$

where $W_N = e^{-j\frac{2\pi}{N}}$.

For a sequence of length N, Equation 2.37 requires N^2 complex products and $N(N-1)$ complex sums. Summing two complex numbers requires summing the respective real and imaginary parts, and thus it requires two real sums. The product of two complex numbers $X_1 = a_1 + jb_1, X_2 = a_2 + jb_2$ instead requires four real products because it is expressed as $X_1 \cdot X_2 = a_1 a_2 - b_1 b_2 + j(a_1 b_2 + b_1 a_2)$.

The computational cost is thus of the order of N^2, because for increasing N, the cost of computing the N^2 products prevails over the $N(N-1)$ sums.[15]

The computational cost of the DFT can be improved by exploiting some of its properties to factor the number of operations. One famous algorithm for a fast Fourier transform (FFT) was devised by Cooley and Tukey (1965). From that time, the FFT acronym has been widely misused instead of DFT. It should be noted that the DFT is the transform, while the FFT is just a fast implementation of the DFT, and there is no difference between the two but the computational cost. The computational savings of the FFT are obtained by noting that it can be written as:

$$X[k] = \sum_{m=0}^{N/2-1} x[2m] W_{N/2}^{km} + W_N^k \sum_{m=0}^{N/2-1} x[2m+1] W_{N/2}^{km} \tag{2.38}$$

thus reducing the DFT to two smaller DFTs of length $N/2$. The coefficients computed by these two DFTs are used as partial results for the computation of two of the final DFT bins. If $N/2$ is even, these two DFTs can be split in two smaller ones each. This procedure can be iterated. If N is a power of 2, the procedure can be repeated up to a last iteration where we have several DFTs of length 2, called *butterflies*. If, for example, $N = 2^M$, there are M stages and for each stage only $N/2$ butterflies need be computed. Since each butterfly has one complex product and two complex sums, we have in total $MN/2$ complex products and MN complex sums, yielding an order of $N\log_2 N$ operations. To evaluate the advantage of the FFT, let us consider $N = 1024$ bins. The DFT requires approximately 1024×1024 operations, while the FFT requires 10×1024 operations – a saving of 100 times (i.e. two orders of magnitude)!

2.4.6 Short-Time Fourier Transform

For long signals, it is hard to compute a single DFT, even when using FFT algorithms that reduce the computational cost. A different approach consists of taking slices of the signal and analyzing them with a DFT. This is particularly useful for a different class of signals, those that are non-stationary. Those signals change their properties (e.g. their frequency content with time), and thus slicing them in small pieces allows you to evaluate how the frequency content changes. The slices will be of the same length. Usually, before applying the DFT, we also apply *windowing* to the slice, which smooths the corners at the beginning and at the end of the slice. In this context, a windowed signal is intended as a slice of the signal multiplied sample by sample with a window of the same length. The window may have a smooth transition at the

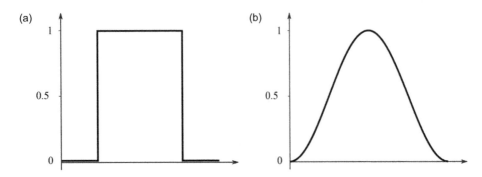

Figure 2.11: A rectangular window (a) and a Hanning window (b).

beginning and end, and unitary amplitude in the middle (see Figure 2.11b). Windowing consists of multiplying sample-wise the slice with the window, thus applying a short fade-in and fade-out to the signal to be analyzed. It can be shown that multiplying a signal with a window alters its frequency content. Please note that even if you do not apply windowing explicitly, slicing a signal is still equivalent to multiplying the original signal by a rectangular window (Figure 2.11a). Thus, altering the signal is unavoidable in any case.[16]

The shape of the window determines its mathematical properties, and thus its effect on the windowed signal. Indeed, windows are formulated in order to maximize certain criteria. In general, there is no optimal window, and selection depends on the task.

To recap what we have discovered in this section:

- Any infinite-length discrete-time signal, or any finite-length signal that is too long to be practical to analyze in its entirety, can be windowed to obtain a finite-length signal, in order to apply the DFT. If the signal is non-stationary, the DFT will highlight the spectral information of the portion of signal we have windowed.
- The operation of windowing alters the content of the signal.

2.5 Once Again on LTI Systems: Filters

After discussing LTI systems and the frequency domain, we can describe filters. Filters are LTI systems that are designed to modify the frequency content of an input signal by cutting or boosting selected components. Common types of filters, defined by their frequency mask, are:

- *Low-pass filters (LPFs)*. These ideally cancel all content above a so-called *cutoff* frequency. In reality, there is a transition band where the frequency components are progressively attenuated. The steepness of the filter in the transition band is also called *roll-off*. The roll-off may end, reaching a floor that is non-zero. This region, called stop-band, has a very high, but not infinite, attenuation. Synthesizer filters usually have a roll-off of 12 or 24 dB/oct, and generally have a small resonant bell at the cutoff frequency.

- *High-pass filters (HPFs).* These ideally cancel those frequency components below a cutoff frequency. They are thus considered the dual of low-pass filters. Also, HPFs have a roll-off and may have a resonance at the cutoff frequency.

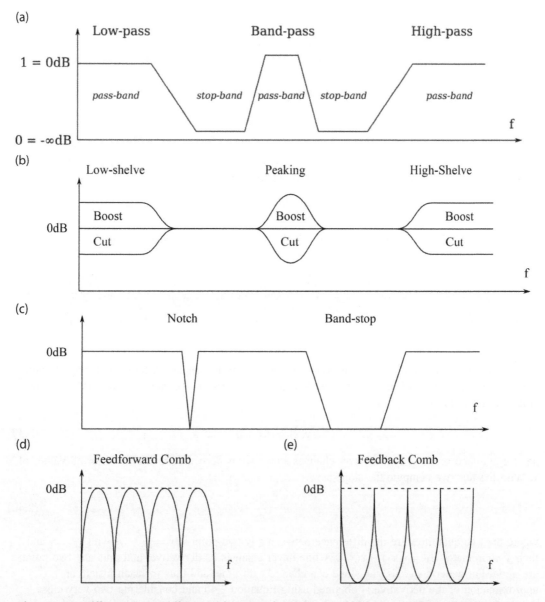

Figure 2.12: Different types of filters: (a) low-pass, band-pass, and high-pass, with pass band and stop band as indicated in the figure, and transition bands in between; (b) shelving and peaking filters; (c) notch and band-stop filters; (d) feedforward comb; and (e) feedback comb.

- *Band-pass filters (BPFs).* These are filters that only select a certain frequency band by attenuating both the low- and the high-frequency content. They are defined by their bandwidth (or its inverse, the quality factor, Q) and their central frequency. They have two transition bands and two stop bands.
- *Shelving filters.* These are filters that cut or boost the content in the bass or treble range, and leave the rest unmodified. They are used in equalizers to treat the two extremes of the audible range. They are defined by the cutoff, the roll-off, and the gain.
- *Peaking filters.* These cut or boost a specific bandwidth, and leave the rest unmodified. They are defined by the bandwidth (or the quality factor, Q), the central frequency, and the gain. They are used in parametric equalizers.
- *Comb filters.* These filters apply a comb-like pattern that alters (cuts or boosts) the spectrum at periodic frequency intervals. They can have feedback (boost) or feedforward (cut) configurations.

Another class of filters, widely employed in DAWs, the so-called *brickwall* filters, are just low-pass or high-pass filters with a very steep roll-off. They are used in audio track mastering applications to remove any content above or below a certain cutoff frequency.

Most synthesizer filters are of the IIR type and have a gentle roll-off (Figure 2.12).

2.6 Special LTI Systems: Discrete-Time Differentiator and Integrator

 TIP: This is the 101 for virtual analog modeling of electronic circuits. Read carefully if you want to get started on this topic.

Later it will be useful to obtain the discrete-time derivative of a signal. A bit of high school math will refresh the concept of derivative. The derivative of a curve is the slope of the curve in a given point. A line of the form

$$y(x) = mx + q \tag{2.39}$$

has slope m and offset q. A line never changes slope, so its derivative is $\dot{y}(x) = m$ for all values of x. With the line, we compute the slope as:

$$m = \frac{\Delta y}{\Delta x} \tag{2.40}$$

where the two quantities are the difference between any two points $\Delta x = x_2 - x_1$ with $x_2 > x_1$ and their y coordinates $\Delta y = y(x_2) - y(x_1)$. A line never changes its derivative, and thus any two points are good. For generic signals, the slope can change with time, and thus for each instant an approximation of the derivative is obtained using Equation 2.40 and considering two very close points. In the case of discrete-time signals, the choice of the points is pretty obvious:[17] there are no two closer points than two consecutive samples $y[n], y[n-1]$. The difference equation for the first-order backward differentiator is thus expressed as:

$$m[n] = y[n] - y[n-1] \tag{2.41}$$

The quantity Δx is the time corresponding to one sample. When transferring a continuous-time problem to the discrete-time domain, therefore, $\Delta x = T_s$. Otherwise, the term can be expressed in terms of samples, thus $\Delta x = 1$. Many DSP books adopt this rule, while physical modeling texts and numerical analysis books use the former to retain the relation between timescales in the two domains.

The frequency response of the ideal differentiator is a ramp rising by 6 dB per octave, meaning that for each doubling of the frequency there is a doubling of the output value. It also means that any constant term (e.g. an offset in the signal) is canceled because it lies at null frequency, as we would expect from a differentiator, that calculates only the difference between pairs of values. It must be said, for completeness, that the digital differentiator of Equation 2.41 slightly deviates at high frequencies from the behavior of an ideal differentiator. Nonetheless, it is a widely used approximation of the differentiation operator in the discrete-time domain. More complex differentiation schemes exist but are extremely complex and bear additional issues, and thus they have very limited application to music signal processing.

A very different case is that of digital integrators, where several approximations exist and are selected depending on the use case. Let us first consider the integration operation. In the continuous-time domain, the integral is the area under the curve corresponding to a signal. In analog electronics, this is done by using an operational amplifier (op-amp) with a feedback capacitor, allowing you to cumulate the signal amplitude over time. Similarly, in the discrete-time domain, it is sufficient to indefinitely cumulate the value of the incoming signal. Similar to the digital differentiator, the digital integrator is just an approximation of the ideal integrator. Two extremely simple forms exist, the forward and the backward Euler – or rectangular – integrators, described by Equations 2.42 and 2.43:

$$y[n] = y[n-1] + T_s x[n] \quad \text{Backward Euler Integrator} \tag{2.42}$$

$$y[n] = y[n-1] + T_s x[n-1] \quad \text{Forward Euler Integrator} \tag{2.43}$$

The forward Euler integrator requires two memory elements but features otherwise similar characteristics. If you want proof that the difference equations (Equations 2.42 and 2.43) implement an integrator, consider this: integrating a curve, or a function, by definition, implies evaluating the underlying area. Our input samples tell us the shape of the curve at successive discrete points. We can take many rectangles that approximate the area between each two consecutive points of that curve. Figure 2.13, for example, shows two rectangles approximating the area under the curve (black line), of which we know three points. The area of these two rectangles slightly underestimates the real value of the area under the curve. Intuitively, by fitting this curve with more rectangles (i.e. reducing their width and increasing the number of points, that is, the sampling rate) the error reduces.

A computer can thus calculate an approximation of the area by summing those rectangles, as follows: the distance between two of them is the sampling interval T_s. This gives us the width of each rectangle. We only have to decide the height of the rectangle. We can take either $x[n]$ or $x[n+1]$, the former for the forward Euler integrator and the latter for the backward integrator. The integrator used in Figure 2.13 is the forward Euler, as the height of the rectangle is taken from $x[n-1]$.

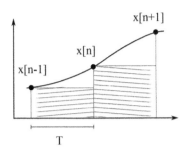

Figure 2.13: A discrete-time feedback integrator applied to a signal. The hatched area results from the integration process, while the remaining white area under the signal (solid bold line) is the approximation error. The shorter the time step, the lower the error.

To obtain the area in real time, as the new samples come in, we don't have to sum all the rectangles at each time step. We can just cumulate the area of one rectangle at a time and store the value in the variable $y[n-1]$. At the next step, we will add the area of a new rectangle to the cumulative value of all previous ones, and so forth.

Other integrators exist that have superior performance but higher computational cost. In this context, we are not interested in developing the theory further, but the reader can refer to Lindquist (1989) for further details.

2.7 Analog to Digital and Back

Discrete-time signals can be directly generated by computers, processed, and plotted on a screen. However, if we deal with music, we want to record them and listen to them!

These two operations are done through two different processes, the analog-to-digital and the digital-to-analog conversions, which are the inverse of each other. Before putting it in mathematical terms, let us provide some background and clarify the terms used in this section. The analog-to-digital conversion of a signal comprises two fundamental steps: *sampling* and *quantization*, in the correct order (Figure 2.14). The first step yields a discrete-time signal, while the second transforms it into a proper digital signal (i.e. a signal described with a discrete and finite set of numbers). This signal is said to have finite precision. If this process is not properly done, it leads to aliasing, an ugly beast that we shall discuss in Section 2.9.

The step of sampling is conducted by taking quick snapshots of a signal at equally spaced time intervals. Let $x_c(t)$ be a continuous-time signal that has a finite bandwidth (i.e. its spectrum is zero – or at least almost inaudible – over a certain frequency F_N). To sample this signal, it can theoretically be multiplied (modulated) by a train of Dirac pulses:

$$s(t) = \sum_{n=-\infty}^{n=\infty} \delta(t - nT_s) \tag{2.44}$$

where the time interval T_s is the inverse of the sampling frequency (or sampling rate, or sample rate) $F_s = \frac{1}{T_s}$. The operation is thus:

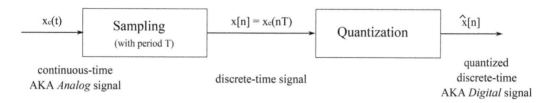

Figure 2.14: The analog-to-digital conversion mechanism (A/D or ADC).

$$x_s(t) = x_c(t) \cdot s(t) = x_c(t) \cdot \sum_{n=-\infty}^{+\infty} \delta(t - nT). \tag{2.45}$$

The result of this operation is a signal that is zero everywhere besides the instants $x_c(nT_s)$. These instantaneous values are stored, using a suitable method, in the vector $x[n] = x_c(nT_s)$.

Is there any particular choice for the F_s? Yes, there is! Let us take a look at $x_s(t)$ in the frequency domain. By evaluating the continuous-time Fourier transform of $s(t)$ and applying its properties, it can be shown that the sampled signal $X_s(f)$ is periodic in the frequency domain, as shown in Figure 2.15. The product of $x_c(t)$ and $s(t)$ is equivalent to the convolution in the frequency domain (per the convolution property of the DFT). The spectrum of $x_c(t)$ is thus replicated in both the positive and negative frequency axis, and each replica has a distance that is equal to F_s. From this observation, we can deduce that there are values of F_s that are too low and will make the replicas overlap. Is this a problem? Yes, it is! Not convinced? Let us examine what happens next when we convert back the discrete signal into a continuous-time signal.

This process is called *reconstruction* of the continuous-time signal. It creates a continuous-time signal by filling the gaps. First, a continuous time-domain signal $x_s(t)$ is obtained from $x[n]$ by multiplying it with a train of Dirac pulses:

$$x_s(t) = \sum_{n=-\infty}^{+\infty} x[n]\delta(t - nT_s) \tag{2.46}$$

yielding a signal that is zero except from the instants nT_s.

Then a low-pass filter is applied that interpolates between the gaps. Such an ideal low-pass filter is called *reconstruction filter*. Intuitively, the scope of this filter is to filter all the replicas seen in Figure 2.15 and get only the one centered around zero. To get confirmation of this idea, let us consider that in the time-domain, being a low-pass filter, it reacts slowly to abrupt changes, and it will thus smooth the instantaneous peaks of $x_s(t)$ (remember that it is zero besides the instants nT_s) and connect them.

Ideally, this filter should have a flat frequency response up to the cutoff frequency, and then drop instantly to zero. What do you think is the best filter cutoff to separate the spectral replicas of Figure 2.15? Surely $F_s/2$ (i.e. the midpoint between two replicas). The shape of this filter is thus a rectangle with a corner at $F_s/2$. The filter is an LTI system, and thus the output of the filter will be the convolution between $x_s(t)$ and the filter impulse response $h(t)$:

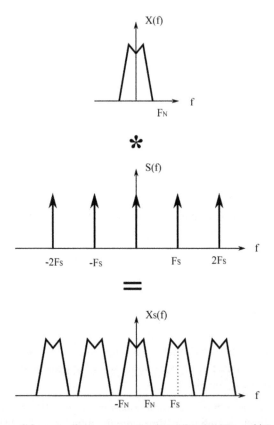

Figure 2.15: The first stage of the sampling process consists of multiplying $x(t)$ by $s(t)$. The product of two signals in the time domain consists of the convolution between their Fourier transform in the frequency domain. The figure shows the two magnitude spectra and the result of the frequency-domain convolution, consisting of replicas of the spectrum of $x(t)$.

$$x_r(t) = x_s(t) * h(t) = x_c(t) \tag{2.47}$$

At this point, the filtered signal is the reconstructed signal we were looking for, and will be exactly the same as the input signal $x_c(t)$. Perfect reconstruction is possible if – and only if – the replicas are at a distance F_s that is large enough to allow the reconstruction filter to cancel all other replicas and their tails. If the replicas do overlap, the filter will let parts of the replicas in, causing aliasing. What we have reported so far is the essence of a pillar theorem, the Nyquist sampling theorem, stating that the sampling rate must be chosen to be *at least* twice the bandwidth of the input signal in order to uniquely determine $x_c(t)$ from the samples $x[n]$. If this condition is true, then the sampling and reconstruction process do not affect the signal at all.

In formal terms, the Nyquist theorem[18] states that if $x_c(t)$ is band-limited:

$$X_c(f) = 0 \quad \text{for} |f| > F_N$$

Table 2.3: DFT of notable signals

Signal	Time Domain	Frequency Domain
Dirac delta	$x[n] = \delta[n]$	$X[k] = 1$
Cosine	$x[n] = \cos(2\pi f n)$	$X[k] = \pi[\delta(k - 2\pi f) + \delta(k + 2\pi f)]$
Rectangular pulse or window	$x[n] = \begin{cases} 1 & -L < n < L \\ 0 & \text{elsewhere} \end{cases}$	$X[k] = L\frac{\sin(kL)}{kL}$
Pulse train	$x[n] = \sum\limits_{m=-\infty}^{m=\infty} \delta(n - mT)$	$X[k] = \sum\limits_{m=-\infty}^{m=\infty} \delta\left(k - \frac{m}{T}\right)$

Note: The figures are only for didactical purposes. They do not consider the effect of sampling that replicates the shown spectra with period F_s. Aliasing will occur for non-band-limited signals.

then $x_c(t)$ is uniquely determined by its samples $x[n] = x_c(nT), n = 0, \pm1, \pm2, \ldots$ if – and only if –

$$F_s > 2F_N. \qquad (2.48)$$

F_N is also called the Nyquist frequency, and is considered the upper frequency limit for a signal sampled at a given F_s.

If Equation 2.48 is not respected, spurious content is added to the original signal that is usually perceived as distortion, or, properly, aliasing distortion.

A question may now arise: If we are not lucky and our signal is not band-limited, how do we deal with it? The only answer is: we band-limit it![19] Any real-world signal that needs to be recorded and sampled is first low-pass filtered with a so-called *anti-aliasing filter*, having cutoff at the Nyquist frequency, thus eliminating any content above it.

A final topic related to filtering: we have seen that the reconstruction filter has its cutoff at the Nyquist frequency, like the anti-aliasing filter, in order to guarantee perfect reconstruction.[20] The idea reconstruction filter is a low-pass filter with cutoff frequency at the Nyquist frequency and vertical slope; in other words, it would look like a rectangle in the frequency domain. From Table 2.3, we observe that its inverse Fourier transform (i.e. the impulse response of this filter) is a symmetrical pulse called *sinc* that goes to zero at infinity, and thus has infinite length. Such an infinite impulse response is hard to obtain. In general, ideal low-pass filters can only be approximated (e.g. by truncating the impulse response up to a certain – finite – length). This makes the reconstruction process not perfect, but trade-offs can be drawn to obtain excellent acoustic results.

2.8 Spectral Content of Typical Oscillator Waveforms

Prior to discussing aliasing, we need to discuss what the spectra of typical oscillator waveforms look like. We shall discuss the following waveforms, easily found on most synthesizers:

- sawtooth;
- triangular; and
- rectangular.

Of course, there are lots of variations thereof, but it is enough to understand the properties of these ones.

Sawtooth signals are usually obtained by a rising ramp that grows linearly and is reset at the end of the period using the modulo operation. Mathematically, it is described as:

$$s[n] = \left(\frac{2}{T}(n \bmod T) \right) - 1 \qquad (2.49)$$

where T is the period of the signal. The signal is divided by T to keep in the range from -1 to 1, and it is shifted by -1 to have zero mean.

The sawtooth spectrum has even and odd harmonics, and the amplitude of each harmonic is dependent of its frequency with a *1-over-f* relation (i.e. with a spectral rolloff of 6 dB/oct). Such a rich spectrum is related to the abrupt jump from 1 to -1. It must be noted that a reversed ramp (i.e. falling) has similar properties.

Triangle waves are, on the other hand, composed by rising and falling ramps, and thus they have no discontinuity. The spectrum of a triangular wave is thus softer than that of a sawtooth wave. Its mathematical description can be derived from that of the sawtooth, by applying an absolute value to obtain the ascending and descending ramps:

$$t[n] = 1 - 2|s[n]| \tag{2.50}$$

The triangle wave has only odd harmonics, and its decay is steeper than that of the sawtooth (12 dB/oct).

The triangle wave has a discontinuity in its first derivative. This can be observed at the end of the ramps: a descending ramp has a negative slope, and thus a negative derivative, while an ascending ramp has a positive slope, and thus a positive derivative. What does the triangle derivative look like? It looks like a square wave! Since the relation between a signal and its derivative, in the frequency domain, is an increase proportional to the frequency, we can easily conclude that the square wave has the same harmonics as the triangle (odd), but with a slower spectral decay (6 dB/oct). The square wave thus sounds brighter than the triangle wave. If we look at the time domain, we can justify the brighter timbre by observing that it has abrupt steps from 1 to −1 that were not present in the triangle wave. The triangular waveform can thus be described as:

$$q[n] = \frac{dr[n]}{dn} = r[n] - r[n-1], \tag{2.51}$$

where we approximated the time derivative with a backward difference as in Equation 2.41. At this point, an important distinction must be made between the square and rectangular waveforms. A square wave is a rectangular waveform with a 50% duty cycle. The duty cycle $\varDelta$ is the ratio between the time the wave is up and the period $\varDelta = T_{on}/T$, as shown in Figure 2.16. This alteration of the wave symmetry also affects the spectral content, and thus the timbre. Remember from Table 2.3 that the DFT of a rectangular window has its zeros at frequencies that depend on the length of the window. Considering that a square wave is a periodic rectangular window, it can be seen as the convolution of a train of Dirac pulses with period T and a rectangular window with T_{ON} equal to half the period T. In the frequency domain, this is equal to the product of the spectra of the window and the pulse train. It can then be shown that the square wave has zeros at all even harmonics. However, if the duty cycle changes, the ratio between the

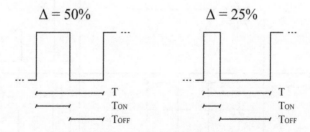

Figure 2.16: Pulse width modulation (PWM) consists of the alteration of the duty cycle (i.e. the ratio between T_{ON} and the period T).

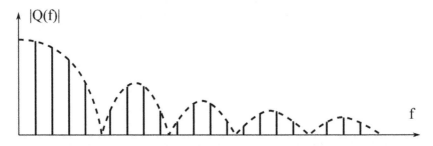

Figure 2.17: A rectangular wave with duty cycle different from 50% has a spectrum that results from the convolution (in time) or the product (in frequency) of the window (a sinc function, dashed lines) and a train of Dirac pulses (producing the harmonics, solid lines).

period and the T_{ON} changes as well, shifting the zeros in frequency and allowing the even harmonics to rise. In this case, the convolution of the Dirac train and the rectangular window will result in lobes that may or may not kill harmonics, but will for sure affect the amplitude of the harmonics, as seen in Figure 2.17. Going toward the limit, with the duty cycle getting close to zero, the result is a train of Dirac pulses, and thus the frequency response gets closer to that of a Dirac pulse train, with the first lobe reaching infinity, and thus all harmonics having the same amplitude. This, however, does not occur in practice.

Table 2.4: Typical oscillator waveforms and related spectra

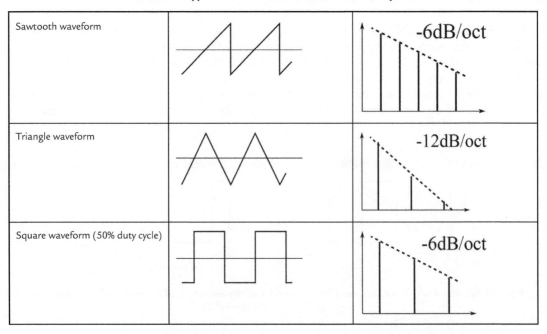

It should be noted that practical oscillator designs do often generate slight variations over the theoretical waveform. An example is provided by the Minimoog Voyager sawtooth that was studied in Pekonen et al. (2011), which shows it to be smoother than the ideal sawtooth wave. A lot of oscillator designs differ from the ideal waveforms described in this section, and a good virtual analog emulation should take this difference into consideration.

2.9 Understanding Aliasing

Since this book is mainly about generating signals inside the computer, we should sit and reflect on this point: If aliasing is only caused by an improper sampling process, why should we care about aliasing in a virtual modular software, where no signal is recorded or sampled from an audio card? Stop and think for a second before moving on.

The answer is: Yes, we still do care about aliasing! In this setting, we do care even more than a studio engineer that records audio signals by digital means. Sound engineers have anti-aliasing filters in their analog-to-digital converters that make any input signal band-limited before sampling. Unfortunately, this is not the case for us software synth freaks, because *generating a discrete-time signal employing a mathematical non-band-limited function is equivalent to sampling that non-band-limited function without an anti-aliasing filter*. In other words, generating a signal (e.g. an ideal sawtooth) in software is equivalent to taking the ideal, non-band-limited continuous-time sawtooth from Plato's Hyperuranion and sampling it. There is no workaround to this, and no anti-aliasing filter can be imposed before this sampling takes place.

Aliasing is one of the most important issues in virtual synthesizers, together with computational cost, and often these two issues conflict with each other: to reduce aliasing, you have to increase the computational cost of the algorithm. The presence of aliasing affects the quality of the signal. Analog synthesizers were not subject to aliasing at all. One notable exception, again, is the Bucket-Brigade delay. BBD circuits are discrete-time systems, and thus subject to the Nyquist sampling theorem. As such, they can generate aliasing. But any other analog gear had no issues of this kind. With the advent of virtual analog in the 1990s, solutions had to be found to generate waveforms without aliasing on the available signal processors of the time. Nowadays, x86 and ARM processors allow a whole lot of flexibility for the developer and improved audio quality for the user.

Common oscillator waveforms (sawtooth, square, and triangle waves) are not band-limited as they have a harmonic content that decays indefinitely at a rate of 6 or 12 db/oct typically, and thus they go well above the 20 kHz limit (ask your dog). This does not pose problems in a recording setting. Any analog-to-digital converter applies an anti-aliasing filter to suppress any possible component above F_N, thus limiting the bandwidth of the signal. Any digital audio content, therefore, if recorded and sampled properly, does not exhibit noticeable aliasing. Unfortunately, this is not the case, because, as we said, generating a discrete-time signal employing a non-band-limited function is equivalent to sampling that non-band-limited function (i.e. freezing its theoretical behavior at discrete points in time, exactly as the sampling process does).

The outcome of an improper sampling process or the discretization of a non-band-limited signal results in the leak of spurious content in the audible range, which is undesired (unless you are producing dubstep, but that is another story). With periodic signals, such as a sawtooth wave sweeping up in frequency, the visible and audible effect of aliasing is the mirroring of harmonics approaching the Nyquist frequency and being reflected to the $0 - F_N$ range. As aliasing gets even more severe, other replicas of the spectrum get in the range $0 - F_N$ and the visible effect is the mirroring of (already mirrored) harmonics to the left bound of the $0 - F_N$ range.

After aliasing has occurred, there is generally no way to repair it, since the proper spectrum and the overlapping spectrum are embedded together.

Fortunately, the scientific literature is rich with methods to overcome aliasing in virtual synthesizers. In Sections 8.2 and 8.3, a couple of methods are exposed to deal with aliasing in oscillators and waveshapers, while other methods are referenced in Section 11.1 for further reading.

Let us examine a signal in the frequency domain. In Section 2.7, we described the sampling process and discovered that it necessarily generates aliases of the original spectrum, but these can be separated by a low-pass filter. The necessary condition is that these are well separated. However, if the original spectrum gets over the Nyquist frequency, the aliases start overlapping with each other, as shown in Figure 2.18. Zooming in, and considering a periodic signal, as in Figure 2.19, with equally spaced harmonics, you can see that after the reconstruction filter, the harmonics of the right alias gets in the way. This looks like the harmonics of the original signal are mirrored. The aliasing components mirrored to the right come, in reality, from the right alias, while those mirrored at 0 Hz come from the left alias.

Let us consider aliasing in the time domain and take a sinusoidal signal. If the frequency of the sine is much lower than the sampling rate (Figure 2.20a), each period of the sine is described by a large number of discrete samples. If the frequency of the sine gets as high as the Nyquist frequency (Figure 2.20b), we have only two samples to describe a period, but that is enough (under ideal circumstances). When, however, the frequency of the sine gets even higher, the sampling process is "deceived." This is clear after the reconstruction stage (i.e. after the reconstruction filter tries to fill in the gaps between the samples). Intuitively, since the filter is a low-pass with cutoff at the Nyquist

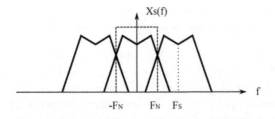

Figure 2.18: Overlapping of aliases, occurring due to an insufficient sampling frequency for a given broad-band signal. The ideal reconstruction filter is shown (dashed, from $-F_N$ to F_N).

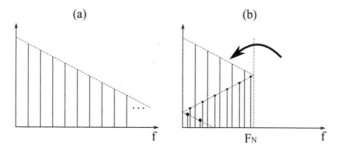

Figure 2.19: The effect of aliasing on the spectrum of a periodic signal. The ideal spectrum of a continuous-time periodic signal is shown. The signal is not band-limited (i.e. sawtooth). When the signal is sampled without band-limiting, the original spectrum, aliases gets in the way. The effect looks like a mirroring of component over the Nyquist frequency (highlighted with a triangle) and a further mirroring at 0 Hz (highlighted with a diamond), producing the zigzag (dashed line). Please note that this "mirroring" in reality comes from the right and left aliases.

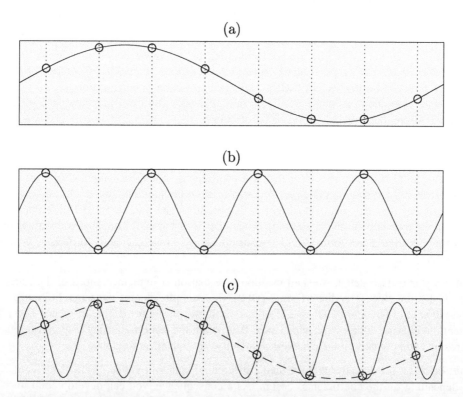

Figure 2.20: Effect of aliasing for sinusoidal signals. (a) A continuous-time sine of frequency $F_N/4$ is shown to be sampled at periodic intervals, denoted by the vertical grid with values highlighted by the circles. (b) A sine at exactly F_N is still reconstructed correctly with two points per period. (c) A continuous-time sine at frequency $(7/4)F_N$ is mistaken as a sine of frequency $F_N/4$, showing the effect of aliasing. The dashed curve shows the reconstruction that is done after sampling, very similar to the signal in (a). In other words, the sine is "mirrored."

frequency, the only thing it can do is fill the gaps (or join the dots) to describe a sine at a frequency below Nyquist (Figure 2.20c). As you can see, when aliasing occurs, a signal is *fooling* the sampling process and is taken for another signal.

2.10 Filters: The Practical Side

Common types of filters were discussed in Section 2.5. Filters are implemented in the discrete domain by unidirectional signal-flow graphs implementing the differential equation that yield the impulse response of the desired filter. A conceptual flow for a filter implementation is the following:

1. Define a frequency response in the frequency domain using a mask.
2. Find a suitable differential equation that implements the filter. These are available from textbooks.
3. Find a suitable design strategy to compute the actual coefficients that yield the desired frequency response.
4. Implement the filter as a flow graph.
5. Translate the flow graph in code (e.g. C/C++).
6. Test the filter (e.g. with a Dirac pulse) to verify that the Fourier transform of the impulse response is approximately the desired one.

In this book, we will not cover filter design techniques, but provide a couple of examples easily implemented in Rack, leaving the reader to specific books. A very important filter topology is the *biquad* filter, or second-order section (SOS). This has application in equalization, modal synthesis, and in the implementation of several IIR filters of higher order, by cascading several SOS. The discrete-time differential equation of the biquad follows:

$$y[n] = b_0x[n] + b_1x[n-1] + b_2x[n-2] - a_1y[n-1] - a_2y[n-2] \qquad (2.52)$$

From its differential equation, it is clear that five coefficients should be computed (we are not covering here how) and four variables are required to store the previous two inputs and previous two outputs. The computational cost of this filter is five products and four sums. The differential equation can be translated in the signal-flow graph shown in Figure 2.21a. This is a realization of the filter that is called direct form 1 (DF1) because it directly implements the difference equation (Equation 2.52).

Without going into deep detail, we shall mention here that other implementations of the same difference equation exist. The direct form 2 (DF2), for example, obtains the same difference equation but saves two memory storage locations, as shown by Figure 2.21b. In other words, it is equivalent but cheaper. Other realizations exist that obtain the same frequency response but have other pros and cons regarding their numerical stability and quantization noise.

It should be clear to the reader that the same differential equation yields different kinds of filters, depending on the coefficients' design. As a consequence, you can develop code for a generic *second-order filter* and change its response (low-pass, high-pass, etc.) or cutoff frequency just by changing its coefficients. If we want to do this in C++, we can thus define a class similar to the following:

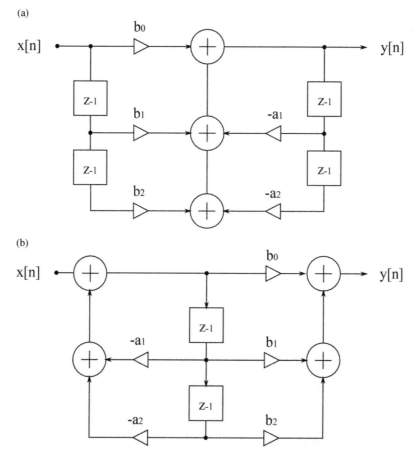

Figure 2.21: Signal-flow graph of a biquad filter in its direct form 1 (DF1) realization (a) and in its direct form 2 (DF2) realization (b).

```
class myFilter {
private:
    float b0, b1, ...; // coefficients
    float xMem1, xMem2, yMem1, ...; // memory elements

public:
    void setCoefficients(int type, float cutoff);
    float process(float in);
}
```

In this template class, we have defined a public function, setCoefficients, that can be called to compute the coefficients according to some design strategy (e.g. by giving a type and a cutoff). The coefficients cannot be directly modified by other classes. This protects them from unwanted errors or bugs that may break the filter integrity (even small changes to the coefficients can cause instability). The `process` function is called to process an input sample and returns an output sample. An implementation of the DF1 SOS would be:

```
float process (float in) {
    float cumulate;
    float out;

    sums = b2 * xMem2 - a2 * yMem2;
    sums += b1 * xMem1 - a1 * yMem1;
    out = b0 * in + sums;
    xMem2 = xMem1;
    yMem2 = yMem1;
    xMem1 = in;
    yMem1 = out;
}
```

As you can see, the input and output values are stored in the first memory elements, while the first memory elements are propagated to the second memory elements.

Please note that, usually, audio processing is done with buffers to reduce the overhead of calling the process function once for each sample. A more common process function would thus take an input buffer pointer, an output buffer pointer, and the length of the buffers:

```
void process(float *in, float *out, int length);
```

As we shall see, this is not the case with VCV Rack, which works sample by sample. The reasons behind this will be clarified in Section 4.1.1, and it greatly simplifies the task of coding.

2.11 Nonlinear Processing

Although the theory developed up to this point is mainly devoted to linear systems, some examples of nonlinear processing are provided due to their relevance in the audio processing field.

2.11.1 Waveshaping

Waveshaping is the process of modifying the appearance of a wave in the time domain in order to alter its spectral content. This is a nonlinear process, and as such it generates novel partials in the signal, hence the timbre modification. The effect of the nonlinearity is evaluated by using a pure sine as input and evaluating the number and level of new partials added to the output. The two most important waveshaping effects are distortion effects and wavefolding, or foldback.

For foldback, we have an entire section that discusses its implementation in Rack (see Section 8.3). We shall now spend a few words on the basics of waveshaping and distortion. When waveshaping is defined as a *static nonlinear function*, the signal is supposed to *enter* the nonlinear function and read the corresponding value. If x is the value of the input signal at a given time instant, the output of the nonlinear function is just $f(x)$. Figure 2.22 shows this with a generic example.

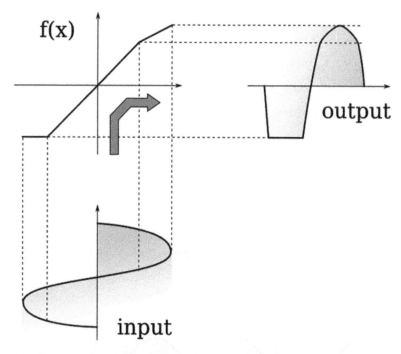

Figure 2.22: A static nonlinear function affecting an input signal. As the term "waveshaper" implies, the output wave is modified in its shape by the nonlinear function. It should be clear to the reader that changing the amplitude of the input signal has drastic effects. In this case, a gain factor that reduces the input signal amplitude below the knees of $f(x)$ will make the output wave unaltered.

As you can see, the waveshaping (i.e. the distortion applied to the signal) simply follows the application of the mapping from x to y. In Figure 2.23, we show how a sine wave is affected by a rectifier. A rectifier is a circuit that computes the absolute value of the input, and it can be implemented by diodes in the hardware realm. The input sine "enters" the nonlinear function, and for each input value we compute the output value by visually inspecting the nonlinear function. As an example, the input value x_1 is mapped to the output value y_1 and similarly for other selected points.

Since the wave shape is very important in determining the signal timbre, any factor affecting the shape is important. As an example, adding a small offset to the input signal, or inversely to the nonlinear function, affects the timbre of the signal. Let us take the rectifier and add an offset. This affects the input sine wave, as shown in Figure 2.24. The nonlinear function is now $y = |x - a|$. You can easily show yourself, by sketching on paper, that the output is equivalent to adding an offset to the sine input and using the previous nonlinearity $y = |x|$.

A static nonlinearity is said to be memoryless. A system that is composed of one or more static nonlinear functions and one or more linear systems (filters) is said to be nonlinear and dynamical, since the memory elements (in the discrete-time domain, the filter states) make the behavior of the system dependent not only on the input, but also on the previous values (its history). We are not going to deal with these systems due to the complexity of the topic. The user should know, however,

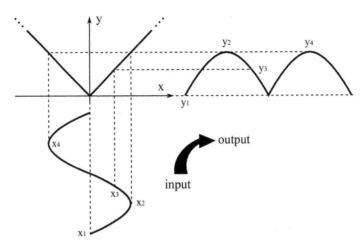

Figure 2.23: Rectification of a sine wave by computing the function $y=|x|$ (absolute value).

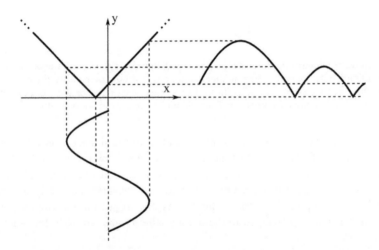

Figure 2.24: Rectification of a sine wave by computing the function $y = |x - a|$. The same result is obtained by offsetting the input waveform by a before feeding it to the nonlinearity.

that there are several ways to model these systems. One approach is to use a series of functions (e.g. Volterra series) (Schetzen, 1980).[21] Another one is to use the so-called Hammerstein, Wiener, or Hammerstein-Wiener models. These models are composed by the cascade of a linear filter and a nonlinear function (Wiener model), or a nonlinear function and a linear filter (Hammerstein model), or the cascade of a filter, a nonlinearity, and another filter (Hammerstein-Wiener model) (Narendra and Gallman, 1966; Wiener, 1942; Oliver, 2001). Clearly, the position of the linear filters is crucial when there is a nonlinearity in the cascade. Two linear filters can be put in any order thanks to the commutative property of linear systems, but when there is a nonlinear component in the cascade this does not hold true anymore. There are many techniques to estimate the parameters of both the linear parts and the nonlinear function, in order to match the behavior of a desired nonlinear system (Wills et al., 2013), which, however, requires some expertise in the field.

Back to our static nonlinearities, let us discuss distortion. Distortion implies saturating a signal when it gets close to some threshold value. The simplest way to do this is *clipping* the signal to the threshold value when this is reached or surpassed. The clipping function is thus defined as:

$$f(x) = \begin{cases} \tau & x \geq \tau \\ -\tau & x \leq -\tau \\ x \text{ elsewhere} \end{cases} \tag{2.53}$$

where τ is a threshold value. You surely have heard of clipping in a digital signal path, such as the one in a digital audio workstation (DAW). Undesired clipping happens in a DAW (or in any other digital path) when the audio level is so high that the digits are insufficient to represent digital value over a certain threshold. This happens easily with integer representations, while floating-point representations are much less prone to this issue.[22] Clipping has a very harsh sound and most of the times it is undesired. Nicer forms of distortion are still S-shaped, but with smooth corners. An example is a soft saturation nonlinearity of the form:

$$f(x) = \tanh(x) \tag{2.54}$$

Another saturating nonlinearity is:

$$f(x) = \text{sign}(x)\left(1 - e^{-|x|}\right) \tag{2.55}$$

Other distortion functions may be described by polynomials such as the sum of the terms:

$$f(x) = x + a_{2x}^2 + a_{3x}^3 + \cdots = \sum_{p=1}^{N} a_p x^p, \tag{2.56}$$

where $a_1 = 1$. This equation generates a signal with second harmonic with amplitude a_2, third harmonic with amplitude a_3, and so forth, up to the nth harmonic. It is important to note that even harmonics and odd harmonics sound radically different to the human ear. The presence of even harmonics is generally considered to add warmth to the sound. A nonlinear function that is even only introduces even harmonics:

$$f(x) = f(-x) \tag{2.57}$$

An odd function only introduces odd harmonics:

$$f(x) = -f(-x) \tag{2.58}$$

There are a number of well-known distortion functions, and many are described by functions that are hard to compute in real time. In such cases, lookup tables (LUTs) may be employed, which help to reduce the cost. This technique is described in Section 10.2.1.

An important parameter to modify the timbre in distortion effects is the input gain, or drive. This gain greatly affects the timbre. Consider the clipping function of Equation 2.53, shown in Figure 2.25a. When the input level is low, the output is exactly the same as the input. However, by rising the input gain, the signal reaches the threshold and gets clipped, resulting in the clipping distortion. Other distortion functions

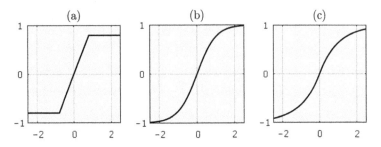

Figure 2.25: Several distortion curves. The clipping function of Equation 2.53 (a), the hyperbolic tangent function (b), and the saturation of Equation 2.55.

behave similarly, with just a smoother introduction of the distortion effect. The smaller the signal, the more linear the behavior (indeed, the function in Figure 2.25b is almost a line in the surroundings of the origin). Again, if we put the drive gain *after* the nonlinear function, its effect is a mere amplification and it does not affect the timbre of the signal that comes out of the nonlinear function.

2.11.2 Amplitude and Ring Modulation

Two common forms of nonlinear processing used in the heyday of analog signal processing are the amplitude and ring modulation techniques. While amplitude modulation (AM) rings a bell to most people because it was the first transmission technology for radio broadcast, and is still used in some countries, the name ring modulation (RM) is of interest only to synthesizer fans. In essence, they are almost similar, and we shall discover – in short – why.

Amplitude modulation as a generic term is the process of imposing the amplitude of a *modulating* signal onto a *carrier* signal. Mathematically, this is simply obtained by multiplying the two. With electronic circuits, this is much more difficult to obtain. The multiplication operation in the analog domain is achieved by a so-called *mixer*.[23] One way to obtain amplitude modulation is to use diode circuits, in particular a diode ring, hence the name ring modulation. The use of diode ring circuits allowed for a form of amplitude modulation that is often called *balanced*, and it differs from the *unbalanced* amplitude modulation for the absence of the carrier in the output. To be more rigorous, the diode ring implements a double-sideband suppressed-carrier AM transmission (DSB-SC AM), while conventional AM contains the carrier (DSB AM). We say "conventional" with reference to DSB AM because it has been widely adopted for radio and TV broadcast. The reason for that is the presence of the carrier frequency in the modulated signal spectrum, which helps the receiver to recover the modulating signal. This allows you to design extremely simple circuits for the receiver. By the way, the envelope follower of Section 7.3 models a simple diode-capacitor circuit to recover the envelope from a modulated signal that is also perfectly suited to demodulate DSB AM signals.

Let us now introduce some math to discuss the spectrum of the modulated signal. The carrier signal $c(t)$ is by design a high-frequency sinusoidal signal $c(t) = A_c \cos(2\pi f_c t + \phi_c)$. In radio broadcast, it has a frequency of the order of tens or hundreds of kHz, while in synthesizer application it is in the audible range. The modulating signal $m(t)$ can be anything from the human voice, to music, and so on.

The DSB-SC, aka RM, simply multiplies the two signals:

$$y(t) = m(t) \cdot c(t) = A_c m(t) \cos(2\pi f_c t + \phi_c) \tag{2.59}$$

Such a process is also called heterodyning.[24] The result in the frequency domain is the shift of the original spectrum, considering the positive and negative frequencies, as shown in Figure 2.26. It can be easily noticed that the modulation of the carrier wave with itself yields a signal at double the frequency $(f_c + f_c)$ and a signal at DC $(f_c - f_c)$. This trick has been used by several effect units to generate an octaver effect. If the input signal, however, is not a pure sine, many more partials will be generated.

The conventional AM differs, as mentioned above, in the presence of the carrier. This is introduced as follows:

$$y(t) = (c(t) + \beta m(t)c(t)) = A_c(1 + \beta m(t)) \cos(2\pi f_c t + \phi_c). \tag{2.60}$$

where, besides the carrier, we introduced the modulation index β to weight the modulating signal differently than the carrier. The result is shown in Figure 2.27.

It must be noted that DSB-SC AM is better suited to musical applications as the presence of the carrier signal may not be aesthetically pleasing. From now on, when referring to amplitude modulation, we shall always refer to DSB-SC AM.

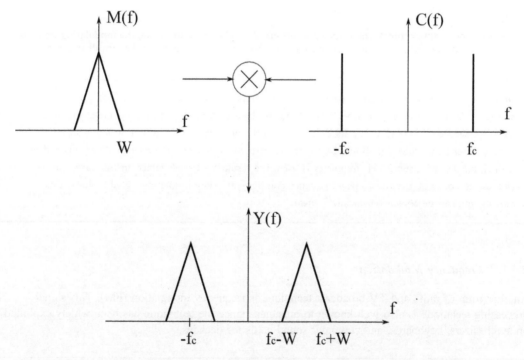

Figure 2.26: The continuous-time Fourier transforms of the modulating and carrier signals, and their product, producing a DSB-SC amplitude modulation, also called ring modulation.

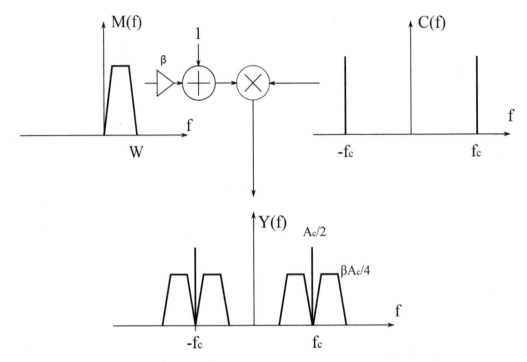

Figure 2.27: Fourier transform of the conventional AM scheme. In this case, the modulating signal is depicted as band-pass signal to emphasize the presence of the carrier in the output spectrum.

In subtractive synthesis, it is common to shape the temporal envelope of a signal with an envelope produced by an envelope generator. This is obtained by a VCA module. In mathematical terms, the VCA is a simple mixer[25] as in ring modulation, with the notable difference that envelope signals should never be negative (or should be clipped to zero when they go below zero). When, instead of an EG, an LFO with sub-20 Hz frequency is used, the result is a *tremolo* effect. In this case, the processing does not give rise to new frequency components. The modulating signal is slowly time-varying and can be considered locally[26] linear.

2.11.3 Frequency Modulation

Another form of radio and TV broadcast technique is frequency modulation (FM). This signal processing technique is also well known to musicians since the technique has been widely exploited on synthesizers, keyboards, and computer sound cards for decades.

The first studies of FM in sound synthesis were conducted by John Chowning, composer and professor at Stanford University. In the 1960s, he started studying frequency modulation for sound synthesis with the rationale that such a simple modulation process between signals in the audible range produce complex and inharmonic[27] timbres, useful to emulate metallic tones such as those of bells, brasses, and so on (Chowning, 1973). He was a percussionist, and was indeed interested in the applications of this technique to his compositions. Among these, *Stria* (1977), commissioned by IRCAM, Paris, is one of the most recognized. At the time, FM synthesis was implemented in digital computers.[28]

In the 1970s, Japanese company Yamaha visited Chowning at Stanford University and started developing the technology for a digital synthesizer based on FM synthesis. The technology was patented by Chowning in 1974 and rights were sold to Yamaha. In the 1980s, the technology was mature enough, and in 1983 the company started selling the most successful FM synthesizer in history, the DX7, with more than 200,000 units sold until production ceased in 1989. Many competing companies wanted to produce an FM synthesizer at the time, but the technology was patented. However, it can be shown that the phase of a signal can be modulated, obtaining similar effects. Phase modulation (PM) was thus used by other companies, leading to identical results, as we shall see later.

The FM technology per se is quite simple and computationally efficient. It consists of modifying the frequency of a carrier signal, say a cosine, by a variable term, that is the instantaneous value taken by a modulating signal. If the carrier wave $c(t) = A_c \cos(2\pi f_c t)$ has a constant frequency f_c, after applying frequency modulation we obtain an instantaneous frequency:

$$f_i(t) = f_c + km(t) \tag{2.61}$$

where k is a constant that determines the amount of frequency skew. At this point, it is important to note that there is a strong relationship between the instantaneous phase and frequency of a periodic signal. Thus, Equation 2.61 can be rewritten as:

$$f_i(t) = f_c + \frac{1}{2\pi} \frac{d\phi(t)}{dt} \tag{2.62}$$

where $\phi(t)$ is the instantaneous phase of the signal. We can thus consider changing the phase of the cosine, instead of its frequency, to obtain the same effect given by Equation 2.61, leading to the following equation:

$$y(t) = A_c \cos(2\pi f_c t + \phi(t)), \tag{2.63}$$

where the phase term is a modulating signal. If it assumes a constant value, the carrier wave is a simple cosine wave and there is no modulation. However, if the phase term is time-varying, we obtain phase modulation (PM). It is thus very important to note that FM and PM are essentially the same thing, and from Equation 2.62 we can state that frequency modulation is the same as a phase modulation where the modulating signal has been first integrated. Similarly, phase modulation is equivalent to frequency modulation where the input signal has been first differentiated.

Let us now consider the outcome of frequency or phase modulation. For simplicity, we shall use the phase modulation notation. Let the modulating signal be $m(t) = \sin(2\pi f_m t)$. The result is a phase modulation of the form:

$$y(t) = A_c \cos(2\pi f_c t + \beta \sin(2\pi f_m t)) \tag{2.64}$$

Considering that the sine is the integral of a cosine, the result of Equation 2.64 is also the same as a frequency modulation with $\cos(2\pi f_m t)$ as the modulating signal.

In the frequency domain, it is not very easy to determine how the spectrum will look. The three impacting factors are f_c, f_m, β. When the modulation index is low (i.e. $\beta \ll 1$), the result is approximately:

$$y(t) = A_c \cos(2\pi f_c t + \beta m(t)) \simeq A_c \cos(2\pi f_c t) - A_c m(t) \sin(2\pi f_c t), \tag{2.65}$$

which states, in other words, that the result is similar to an amplitude modulation with the presence of the carrier (cosine) and the amplitude modulation between the carrier (sine) and the modulating signal. This is called *narrowband FM*. With increasing modulation index, the bandwidth increases, as does the complexity of the timbre. One rule of thumb is Carson's rule, which states that approximately 98% of the total energy is spread in the band:

$$B_c = 2(\beta + 1)W, \tag{2.66}$$

where W is the bandwidth of the modulating signal. This rule is useful for broadcasting, to determine the channels' spacing, but does not say much about the actual spectrum of the resulting signal, an important parameter to determine its sonic result. Chowning studied the relations between the three factors above, and particularly the ratio f_c/f_m, showing that if the ratio is a rational number,[29] the spectrum is harmonic and the fundamental frequency can be computed as $f_0 = f_c/N_1 = f_m/N_2$ if the ratio f_c/f_m can be normalized to ratio of the common factors N_1/N_2. Equation 2.63 describes what is generally called an *operator*,[30] and may or may not have an input $m(t)$. Operators can be summed, modulated (one operator acts as input to a second operator), or fed back (one operator feeds its input with its output), adding complexity to the sound. Furthermore, operators may not always be sinusoidal. For this reason, sound design is not easy on such a synthesizer. The celebrated DX7 timbres, for instance, were obtained in a trial-and-error fashion, and FM synthesizers are not meant for timbre manipulation during a performance as subtractive synthesizers are.

To conclude the section, it must be noted that there are two popular forms of frequency modulation. The one described up to this point is dubbed *linear FM*. Many traditional synthesizers, however, exploit *exponential FM* as a side effect (it comes for free) of their oscillator design. Let us consider a voltage-controlled oscillator (VCO) that uses the V/oct paradigm. Its pitch is determined as:

$$f_0 = f_{ref} \cdot 2^{v_i/12}, \tag{2.67}$$

where the reference frequency f_{ref} may be the frequency of a given base note, and the input voltage v_i is the CV coming from the keyboard. If the keyboard sends, for example, a voltage value of 2 V and $f_{ref} = 440$ Hz, then the obtained tone has $f_0 = 1600$ Hz, two octaves higher than the reference voltage. If, however, we connect a sinusoidal oscillator to the input of the VCO, we modulate the frequency of the VCO with an exponential behavior. The modulation index can be implemented as a gain at the input of the VCO. Please note that if the input signal is a low-frequency signal (e.g. an LFO signal), the result is a vibrato effect.

Exponential FM has some drawbacks with respect to linear FM, such as dependence of the pitch on the modulation index, and is not really used for the creation of synthesis engines. It is rather left as a modulation option on subtractive synthesizers. Indeed, exponential FM is used mostly with LFO. Being slowly time-varying, frequency modulation with a sub-20 Hz signal can be considered locally linear,[31] and there are no novel frequency components in the output, but rather a vibrato effect.

2.12 Random Signals

TIP: Random signals are key elements in modular synthesizers: white noise is a random signal, and is widely used in sound design practice. Generative music is also an important topic and relies on random, probabilistic, pseudo-periodic generation of note events. In this section, we shall introduce some intuitive theory related to random signals, avoiding probability and statistical theoretical frameworks, fascinating but complex.

First of all, what do we mean by random signals?

In general, a discrete-time random sequence $w[n]$ is a sequence with samples generated from a stochastic process (i.e. a process that for all n outputs a value which cannot be predicted).[32] As an example, the tossing of a coin at regular time intervals is a stochastic process that produces random binary values. Measuring the impedance of a number of loudspeakers taken from the end of line of a production facility yields random impedance values that slightly deviate from the expected value. Both processes are stochastic and generate random values. However, the first has a binary outcome and is related to time, while the second produces real values and is not inherently related to time (we suppose all loudspeakers are measured at once).

Important stochastic processes are those related to the Brownian motion of particles, because they take an important role in the generation of noise in electronic circuits or in radio communication, and – as we have discussed in our historical perspective in Chapter 1 – these are the fields that have carried to the development of the signal and circuit theory that is behind synthesizers.

Since current is given by the flow of electrons and these are always randomly moving from atom to atom,[33] there will always be current fluctuations in circuits that produce noise, even if they are not connected to a power supply. When no power supply is connected, the electrons move randomly toward one end of the conductor with the same probability as to the opposite end, and thus in the long run the current is always zero. However, at times when, by chance, more electrons are moving toward one side rather than the other, there will be a slight current flow. At a later moment, those electrons will change their mind and many will turn back, resulting in a current flow of opposite sign. As we said, averaging over a certain time frame, the mean current is zero; however, looking at a short timescale (and with a sensitive instrument), the current has a lot of random fluctuations that can be described statistically. The study of thermal noise is important in electronic and communication circuits design. In audio engineering, for example, amplifiers will amplify thermal noise if not properly designed and cascaded, resulting in poor sound quality. The properties of the noise are important as well, especially for those kinds of noise that do not have equal amplitude over the frequency range.

Back to the theory: if we consider the evolution of the random value in time, we obtain a random signal. With analog circuits, there are tons of different stochastic processes that we can observe to extract a random signal. In a computing environment, however, we cannot toss coins, and it is not convenient at all to dedicate an analog-to-digital converter to sample random noise just to generate a random signal. We have specific algorithms instead that we can run on a computer, each one with its own statistical properties, and each following a different *distribution*.

The distribution of a process is, roughly speaking, the number of times each value appears if we make the process last for an infinite amount of time. Each value thus has a *frequency of occurrence* (not to be confused with the temporal or angular frequency discussed in Section 2.4). In this context, the frequency of occurrence is interpreted as the number of occurrences of that value in a time frame.

Let us consider an 8-bit unsigned integer variable (unsigned char). This takes values from 0 to 255. Let us create a sequence of ten values generated by a random number generator that I will disclose later. We can visualize the distribution of the signal by counting how many times each of the 256 values appears. This is called a *histogram*. The ten-value sequence and its histogram are shown in Figure 2.28a. As you can see, the values span the whole range 0–255, with unpredictable changes. Since we have a very short number of generated values, most values have zero occurrences, some can count one occurrence, and one value has two occurrences, but this is just by chance. If we repeat the generation of numbers, we get a totally different picture (Figure 2.28b). These are, in technical terms, two *realizations* of the same *stochastic process*. Even though both sequences and their histograms are not similar at all, they have been generated using exactly the same algorithm. Why do they look different? Because we did not perform the experiment properly. We need to perform it on a larger run. Figure 2.29 shows two different runs of the same experiment with 10,000 generated values. As you can see, both values tend to be equally frequent, with random fluctuations from value to value. As long as the number of generated values increases, the histogram shows that these small fluctuations get smaller and smaller. With 1 million generated values, this gets clearer (Figure 2.30), and with the number of samples approaching infinity the distribution gets totally flat.

What we have investigated so far is the distribution of the values in the range 0–255. As we have seen, this distribution tends to be *uniform* (i.e. all values have the same probability of occurring). The algorithm used to generate this distribution is an algorithm meant to specifically generate a uniform distribution. Such an algorithm can be thought of as an extraction of a lottery number from a raffle box. At each extraction, the number is put back into the box. The numbers obtained by such an algorithm follow the uniform distribution, since at any time every number has the same chance of being extracted.

Another notable distribution is the so-called *normal* distribution or Gaussian distribution. The loudspeaker samples mentioned above may follow such a distribution, with all loudspeakers having a slight deviation from the expected impedance value (the desired one). The statistical properties of such a population are studied in terms of average μ (or mean) and standard deviation σ. If we model the impedance as a statistic variable following a normal distribution, we can describe its behavior in terms of mean and standard deviation and visualize the distribution as a bell-like function (a Gaussian) that has its peak at the mean value and has width dependent on the standard deviation.

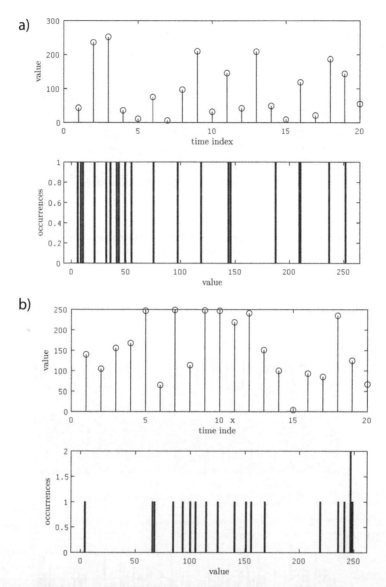

Figure 2.28: Two different sequences of random numbers generated from the same algorithm (top) and their histograms (bottom).

The standard deviation is interpreted as how much the samples deviate from the mean value. If the loudspeakers of the previous example are manufactured properly, the mean is very close to the desired impedance value and the standard deviation is quite small, meaning that very few are far from the desired value. Another way to express the width of the bell is the variance σ^2, which is the square of the variance. To calculate the mean and standard deviation from a finite population (or, equally, from discrete samples of a signal), you can use the following formulae:

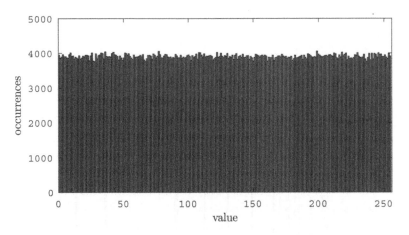

Figure 2.29: Histograms of two random sequences of 10,000 numbers each. Although different, all values tend to have similar occurrences.

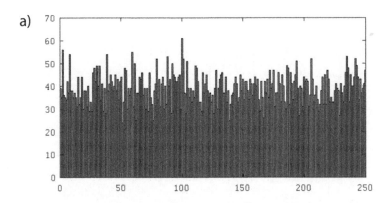

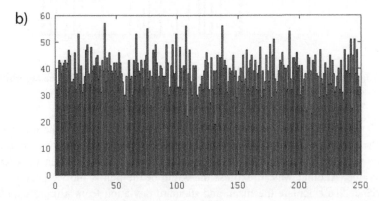

Figure 2.30: Histogram of a sequence of 1 million random numbers. All values tend to have the same occurrences.

$$\mu = \frac{1}{N}\sum_{i=1}^{N} x_i \qquad (2.68)$$

$$\sigma = \sqrt{\frac{1}{N}\sum_{i=1}^{N}(x_i - \mu)^2} \qquad (2.69)$$

Two examples of normal distribution are shown in Figure 2.31.

Going back from loudspeakers to signals, both the normal and uniform random distributions are useful in signal processing, depending on the application. Both uniform and normal random signals are classified as white noise sources, since the spectral content is flat in both cases. The difference between the two is in the distribution of the values (i.e. the amplitude of a signal), with Gaussian noise being more concentrated around the mean.

When we interpret white[34] noise sources in the frequency domain, we have to be careful. These are said to be white, meaning that they have equal energy on all the frequency bands of interest; however, if we calculate the DFT of a window of white noise signal, we will not see a flat spectrum as that of a Dirac pulse. Similar to the frequency of occurrence, which tends to be flat if we take a lot of random samples, the DFT of a white noise window tends to be flat if we take a large number of samples. Similarly, we can average the frequency spectra obtained by several windows of white noise to see that it tends to get flat. Random signals are treated in frequency by replacing the concept of spectrum with the concept of *power spectral density*. We will not need to develop this theory for the purposes of this book; however, if the reader is interested, he or she can refer to Oppenheim and Schafer (2009).

White noise is not colored (i.e. it is flat in frequency). Pink noise is an extremely important form of colored noise. It is very common to find white and noise sources in synthesizers. Pink noise is also named 1-over-f noise, $1/f$ noise, because its power spectral density is inversely proportional to the

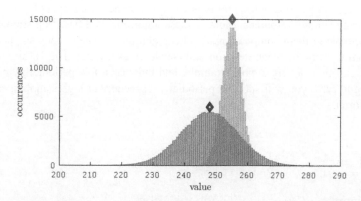

Figure 2.31: Comparison of two normal distributions. The mean values are indicated by the diamond, and are, respectively, 248 and 255. The broader curve has larger variance, while the taller curve has lower variance. If these populations come from two loudspeaker production lines and the desired impedance is 250 Ω, which one of the two would you choose?

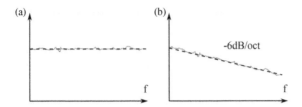

Figure 2.32: Power spectral density of white noise (a) and pink noise (b). The latter exhibits a 6 dB/oct (or 20 dB/decade) spectral decay.

frequency. In other words, it decays in frequency by -20 dB/decade or -6 dB/octave. For this reason, it is darker than white noise and more usable for musical application. Since the slope imposed by integration of a signal is -6 dB/octave, it can also be obtained from white noise by integration.

Having provided a short, intuitive description of random signals, a few words should be dedicated to discrete-time random number generation algorithms. Without the intent of being exhaustive, we shall just introduce a couple of key concepts of random number generators. While natural processes such as thermal noise are totally unpredictable, random number generators are, as any other software algorithm, made by sequences of machine instructions. As such, these algorithms are deterministic (i.e. algorithms that always produce the same outcome given the same initial conditions). This makes them somewhat predictable. For this reason, they are said to be *pseudorandom*, as they appear random on a local scale, but they are generated by a deterministic algorithm.

Thinking in musical terms, we can produce white noise by triggering a white noise sample. However, this sample is finite, and requires looping to make it last longer. Further repetitions of the sample will be predictable as we can store the first playback and compare the new incoming values to the ones we stored, spotting how the next sample will look. The periodicity of the noise sample can be spotted by the human ear if the period is not too long.

There are better ways to produce random numbers that are based on complex algorithms; however, these algorithms are subject to the same problem of periodicity, with the only added value that they do not require memory to store samples, and thus can be extremely long (possibly more than your life) without requiring storage resources. Most of these algorithms are based on simple arithmetic and bit operations, making them computationally cheap and quick. The C code snippet below calculates a random 0/1 value at each iteration and can serve as a taster. The 16-bit variable rval can be accessed bitwise. This is a very simple example, and thus has a low periodicity. Indeed, if you send it to your sound card, you will spot the periodicity, especially at high sampling rates, because the period gets shorter.

```
for (i = 0; ; i++) {
fout = rval.bit0 ^ rval.bit1;
fout ^= rval.bit11 ^ rval.bit13;
rval <<=1;
rval.b0 = fout; // create a feedback loop
// use the random value fout for your scopes
}
```

Random number generators are initialized by a *random seed* (i.e. an initial value – often the current timestamp – that changes from time to time), ensuring the algorithm is not starting every time with the same conditions. If we start the algorithm twice with the same seed, the output will be exactly the same. In general, this is more important for cryptography rather than our musical generation of noise, but keep it in mind, especially during debugging, that random sequences will stay the same if we use the same seed each time, making an experiment based on random value reproducible.

2.13 Numerical Issues and Hints on Coding

Before concluding the boring part of the book, I want to write down a few remarks related to coding for math in C++. The topics we are going to discuss here are:

- computational cost of basic operators;
- issues with division; and
- variable types and effects on numerical precision.

First of all, a few things must be said regarding numerical formats. A lot of DSP literature of the past dealt with fixed-point processing, when floating-point hardware was more expensive. Fixed-point processing makes use of integer numbers to represent fractional numbers using a convention decided by the developer. Fixed-point arithmetic requires a lot of care and is subject to issues of numerical precision and quantization (a topic that we can neglect in this book). Since VCV Rack runs on x86 processors, there is a lot of floating-point power available. We will not even discuss fixed-point arithmetic here, and after all, with floating-point numbers, developing is easier and quicker.

Two types of floating-point variables are available: single precision (32-bit) and double precision (64-bit). The ubiquitous floating-point format is described by IEEE 754, nowadays implemented in most architectures, from personal computers, to smartphones and advanced microcontrollers. The standard describes both single- and double-precision formats. Single-precision floats represent numbers using an 8-bit exponent, a 23-bit mantissa, and a sign bit. They are thus able to span a range from $\pm 3.4 \times 10^{38}$ to $\pm 1.2 \times 10^{-38}$. Impressive, uh? They do so by using a nonuniform spacing, using smaller gaps between adjacent numbers in the low range (toward zero) and larger gaps when going toward infinity. Numbers smaller in magnitude than $\pm 1.2 \times 10^{-38}$ can be represented as well but require special handling. These are called *denormals*, and can slow down computing performances by a factor of even 100 times, so they should be avoided.

Double-precision floating-point numbers are not really required for normal operations in audio DSP; however, there are cases when the double precision helps to avoid noise and other issues. There is a small case study in this book; however, most of the time, single-precision floating-point numbers will be used.

The most common mathematical operators we are going to use in the book are sum, difference, product, division, and conditional operators, as well as modulo. Current x86 processor architectures implement these instructions in several flavors and for both integer and floating-point operands. Each type of instruction takes different clock cycles, depending on its complexity and the type of operands. From the technical documentation of a recent and popular x86 processor (Fog, 2018), one can notice that a sum between two integer operands takes one cycle while the sum of two floating-point operands takes three cycles. Similarly, the product of two integer operands takes three cycles, while with floating-point operands it takes five. The division between two integer operands requires at least 23 cycles, while for floating-point operands only 14 to 16 are required. The division instruction also calculates the remainder (modulo). As you can see, divisions take a time larger by an order of magnitude. For this reason, divisions should be reduced as much as possible, especially in those parts of the code that are executed more often (e.g. at each sample). Sometimes divisions can be precalculated and the result can be employed for the execution of the algorithm. This is usually done automatically by the C++ code compiler if optimizations are enabled. For instance, to reduce the amplitude of a signal by a factor of 10, both the following are possible:

```
signal = signal / 10;
signal = signal * 0.1
```

If compiler optimizations are enabled, the compiler automatically switches from the first to the second in order to avoid division. However, when both operands can change at runtime, some strategies are still viable. Here is an example: if the amplitude of a signal should be reduced by a user-selectable factor, we can call a function only when the user inputs a new value:

```
normalizationFactor = 1 / userFactor;
```

and then use `normalizationFactor` in the code executed for each sample:

```
signal = signal * normalizationFactor;
```

This will save resources because the division is done only once, while a product is done during normal execution of the signal processing routine.

Another issue with division is the possibility of dividing by zero. The result of a division by zero is infinity, which cannot be represented in a numerical format. The value "not a number" (NaN) will appear instead. NaN must be avoided and filtered away at any cost, and division by zero should be avoided by proper coding techniques, such as adding a small non-null number in the denominator of a division. If you develop a filter that may in some circumstances produce NaNs, you should add a check on its output that resets its states and clears the output from these values.

Calculating the modulo is less expensive when the second operand is a power of 2. In this case, bit operations can be used, requiring less clock cycles. For instance, to reset an increasing variable when it reaches a threshold value that is a power of 2 (say 16), it is sufficient to write:

```
var++;
var = var & 0xF;
```

where "0x" is used to denote the hexadecimal notation. 0xF is decimal 15. When var is, say, 10 (0xA, or in binary format 1010), a binary AND with 0xF (binary 1111) will give back number 10. When var is 16 (0x10, binary 10000), the binary AND will give 0.

For what concerns double- and single-precision execution times, in general, double-precision instructions execute slower than single-precision floating-point instructions, but on x86 processors the difference is not very large. To obtain maximum speedup, in any case, it is useful to specify the format of a number to the compiler so that unnecessary operations do not take place. In the following line, for example:

```
float var1, var2;
var2 = 2.1 * var1;
```

the literal 2.1 will be understood by the compiler as a double-precision number. The variable `var1` will thus be converted to double precision, the product will be done as a double-precision product, and finally the result will be converted to single precision for the assignment to `var2`. A lot of unnecessary steps! The solution is to simply state that the literal should be a single-precision floating-point value (unless the other operand is a double and you know that the operation should have a double precision):

```
var2 = 2.1f * var1;
```

The "f" suffix will do that. If you are curious about how compilers transform your code, you can perform comparisons using online tools that are called *compiler explorers* or interactive compilers. These are nice tools to see how your code will probably be compiled if you are curious or you are an expert coder looking for extreme optimizations.

To conclude, let us briefly talk about conditional statements. The use of conditional clauses (`if-else`, '?' operator) is not recommended in those parts of the code that can be highly optimized by the compiler, such as a for loop. When no conditional statements are present in the loop, the compiler can unroll the loop and parallelize it so that it can be executed in a batch fashion. This can be done because the compiler knows that the code will always execute in the same way. However, if a conditional statement is present, the compiler cannot know beforehand how the code will execute, if it will jump and where, and thus the code will not be optimized. In DSP applications, the main reason for performing `for` loops is the fact that the audio is processed in buffers. As we shall see later, in VCV Rack, the audio is processed sample by sample, requiring less use of loops.

2.14 Concluding Remarks

In this chapter, we have examined a lot of different aspects of signals and systems. Important aspects covered along this chapter include the following:

- Continuous-time signals are not computable because they are infinite sets of values and their amplitude belongs to the real set. Discrete-time signals can be computed after discretization of their amplitude value (quantization).
- Discrete-time signals are easy to understand and make some jobs quick. As an example, integral and derivative operations become sum and difference operators, while amplitude modulation, which requires complex electronic circuits, is easily obtained by multiplication.
- Discrete-time systems are used to process discrete-time signals. These systems can be linear and nonlinear. Both are important in musical applications. A review of these systems is done to contextualize the rest of the book.
- The Fourier theory has been reviewed, starting from DFS and getting to the DFT and the STFT. Looking at a signal in the frequency domain allows you to observe new properties of a signal and also allows you to perform some operations faster (e.g. convolution).

- The FFT is just an algorithm to compute the DFT, although these terms are used equivalently in audio processing parlance.
- The sampling theorem is discussed, together with its implications on aliasing and audio quality.
- Ideal periodic signals generated by typical synthesizer oscillators have been discussed.
- Random signals have been introduced with some practical examples. These signals are dealt with differently than other common signals. White noise is an example of a random signal.
- The frequency of occurrence of a random value should not be confused with the angular frequency of a signal.
- Only pseudorandom signals can be generated by computers, but they are sufficient for our applications.
- A few useful tips related to math and C++ are reported.

Exercises and Questions

- Do you produce or record music? If so, your DAW is linear or not? Are there input ranges where we can consider the system linear and ranges where we cannot consider it as linear?
- We know that a distortion effect is non-linear and a filter is linear. What happens if we combine them in a cascade? Does the result change if we put the distortion before or after the filter?
- Suppose you have a spectrum analyzer. You attach a sawtooth oscillator to it and watch its spectral content to reach as high as 24kHz. If you connect the oscillator to a digital effect with a sampling rate of 44.1kHz, and feed the digital effect output to the spectrum analyzer, what will you likely observe? What if the digital effect has a sampling rate of 48kHz?
- Table 2.3 shows the DFTs of notable signals, however it does not consider the effect of the sampling. Are there non-bandlimited signals in the table? If so, how should the DFTs look like in reality?
- Can you devise a nonlinear function that transforms an input sine wave into a square wave? Can you see how an added offset changes it into a rectangle wave with variable duty cycle?

Notes

1 As a student, it bothered me that a *positive* time shift of *T* is obtained by *subtracting T*, but you can easily realize why it is so.
2 Sometimes it is fun to draw connections between signal processing and other scientific fields. Interesting relations between time and frequency exist that can only be explained with quantum mechanics, and carry to a revised version of Heisenberg's uncertainty principle, where time and frequency are *complementary* variables (Gabor, 1946).
3 Nowadays, even inexpensive 8-bit microcontrollers such as Atmel 8-bit AVR, used in Arduino boards, have multiply instructions.
4 *Spoiler alert*: We shall see later that the whole spectrum that we can describe in a discrete-time setting runs from 0 to 2π, and the interesting part of it goes from 0 to π, the upper half being a replica of the lower half.
5 Another tip about scientific intersections: signal processing is a branch of engineering that draws a lot of theory from mathematics and geometry. Transforms are also studied by functional analysis and geometry, because they can be seen in a more abstract way as means to reshape signals, vectors, and manifolds by taking them from a space into another. In signal processing, we are interested in finding good transforms for practical applications or topologies that allow us to determine distances between signals and find how similar they are, but we heavily rely on the genius of mathematicians who are able to treat these high-level concepts without even needing practical examples to visualize them!

6   Different from the transforms that we shall deal with, the human ear maps the input signal into a two-dimensional space with time and frequency as independent variables, where the frequency is approximately logarithmically spaced. Furthermore, the ear has a nonlinear behavior. Psychoacoustic studies give us very complex models of the transform implied by the human ear. Here, we are interested in a simpler and more elegant representation that is guaranteed by the Fourier series and the Fourier transform.

7   We shall see that there are slightly different interpretations of the frequency domain, depending on the transform that is used or whether the input signal is a continuous- or discrete-time one.

8   Following the convention of other textbooks, we shall highlight that a signal is periodic by using the tilde, if this property is relevant for our discussion.

9   Partials may also be inharmonic, as we shall see with non-periodic signals.

10 Engineering textbooks usually employ the letter j for the imaginary part, while high school textbooks usually employ the letter i.

11 The negative frequency components are redundant for us humans, but not for the machine. In general, we cannot just throw these coefficients away.

12 Non-stationary signals cannot be described in the framework of the Fourier transform, but require mixed time-frequency representations such as the short-time Fourier transform (STFT), the wavelet transform, and so on. Historically, one of the first comments regarding the limitations of the continuous-time Fourier transform can be found in Gabor (1946) or Gabor (1947).

13 For the sake of completeness: the DFT also accepts complex signals, but audio signals are always real.

14 Do not try this at home! Read a book on FIR filter design by windowing first.

15 The exact number is $4N^2$ real products and $(N(4N - 2))$ real sums.

16 Unless the original signal is zero everywhere outside the taken slice.

17 Although different approximation schemes exist.

18 It is very interesting to take a look at the original paper from Nyquist (1928), where the scientist developed his theory to find the upper speed limit for transmitting pulses in telegraph lines. These were all but digital. Shannon (1949) provided a proof for the Nyquist theorem, integrating it in his broader information theory, later leading to a revolution in digital communications.

19 More on this in the next section.

20 It actually is the same as the anti-aliasing filter. Symmetries are very common in math.

21 I cannot avoid mentioning the famous Italian mathematician Vito Volterra. He was born in my region, but his fortune lied in having escaped early in his childhood this land of farmers and ill-minded synth manufacturers.

22 The maximum value obtained with a signed integer is $2^{N-1} - 1$. This is 32,767 for $N = 16$ bits, or around 8.3 million for $N = 24$ bits. Floating-point numbers can reach up to $\pm 3.4 \times 10^{38}$.

23 Not to be confused with an audio mixer (i.e. a mixing console). In electronic circuits and communication technology, a mixer is just a multiplier.

24 Incidentally, it should be mentioned that the Theremin exploits the same principle to control the pitch of the tone.

25 Again, here, mixer is not intended as an audio mixing console.

26 The modulating signal can be considered constant for a certain time frame.

27 Inharmonic means that the partials of the sound are not integer multiples of the fundamental, and are thus not harmonic.

28 The original research by Chowning was conducted by implementing FM as a dedicated software on a PDP-1 computer and as a MUSIC V program.

29 The two frequencies are integers.

30 Actually, Equation 2.63 describes a PM operator. An FM operator requires the integral term, resulting in a somewhat more complex equation: $y(t) = A_c \cos\left(2\pi f_c t + 2\pi k \int m(\tau)d\tau\right)$. Yamaha operators also feature an EG and a digitally controlled amplifier, but for simplicity we shall not discuss this.

31 The modulating signal is approximately constant for a certain time frame.

32 Similar considerations can be done for a continuous-time random signal, but dealing with discrete-time signals is easier.

33 Unless we are at absolute zero.

34 The term "white" comes from optics, where white light was soon identified as a radiation with equal energy in all the visible frequency spectrum.

CHAPTER 3

VCV Rack Basics

This chapter provides an overview of VCV Rack. It is not meant as a tutorial on synthesis with Rack, nor as a guide about all available modules. There are hundreds of modules and tens of developers, thus making it barely possible to track all of them down. Furthermore, the development community is very lively and active, and the modules are often updated or redesigned. The chapter just provides a common ground for the rest of the book regarding the platform, the user interface, and how a user expects to interact with it. It shall therefore introduce the terminology and get you started as a regular user, with the aim of helping you design your own modules with an informed view on the Rack user experience.

So, what is VCV Rack exactly? VCV Rack is a standalone software, meant to emulate a modular Eurorack system. It can also host VST instruments employing a paid module called VCV Host. In future releases, a paid VST/AU plugin version of Rack will be available, to be loaded into a DAW. According to VCV founder Andrew Belt, however, the standalone version will always be free of charge. At the moment, Rack can be connected to a DAW using a module called Bridge; however, its support will be discontinued as soon as the VST/AU plugin version is out. To record audio in VCV, the simplest option is to use the VCV Recorder module.

At the first launch, Rack opens up showing a template patch, a simple synthesizer, where modules can be added, removed, and connected. Patches can be saved and loaded, allowing reuse, exchange, and quick switch for performances.

Modules are available from the plugin store. In addition, they can be loaded from a local folder. This is the case for modules that you are developing and are not yet public but you need to test on your machine. Modules can be divided into a few families:

- *Core modules*. Built-in modules that are installed together with Rack, for audio and MIDI connectivity.
- *Fundamental modules*. Basic building blocks provided by VCV for free. These include oscillators, filters, amplifiers, and utilities.
- *VCV modules*. Advanced modules from VCV. These are not free; their price sustains the VCV Rack project.
- *Third-party modules*. These are developed independently by the community and are available under the online plugin store. Some of them are free, some of them are paid, some are open-source, and some are not. If you are reading this book, you are probably willing to build new third-party plugins. Some developers provide both a commercial and a smaller free package (e.g. see Vult Premium and Free plugins, which provide exceptional virtual analog modelling of real hardware circuits).
- *Eurorack replicas*. Some modules available on the store are authorized replicas of existing Eurorack modules. These are free and include, among others, Audible Instrument, Befaco, Synthesis Technology (E-series), and Grayscale modules.

3.1 Overview of the System

Figure 3.1 shows a screen of Rack. Its interface is minimal, with just a few items on the top menu and the rest of the screen free for patching. Audio and MIDI settings, different from other music production software, are not hidden in a menu, but are directly available from the core modules.

The top bar provides access to basic functionalities, including:

- managing patches (New, Open, Save, Save as, Save as template, Revert);
- undo/redo and cable removal;
- view functions (enabling parameter tooltips, zoom, cable opacity and tension, fullscreen mode);
- locking the modules to avoid accidentally dragging out of position;
- visualizing the CPU time consumption of the modules;
- changing sampling rate and thread count;
- managing and updating plugins;
- visualizing the manual, the VCV Rack webpage; and
- opening the user folder.

Modules can be added by right-clicking on an empty area of the rack. This will open the module browser. Modules can be positioned by dragging unless the View → Lock Modules option is enabled. Cables are created by dragging the mouse from an input port to an output port, and vice versa. Right-clicking a module opens up its context menu, while right-clicking a knob or switch allows you to set its value precisely or reset it.

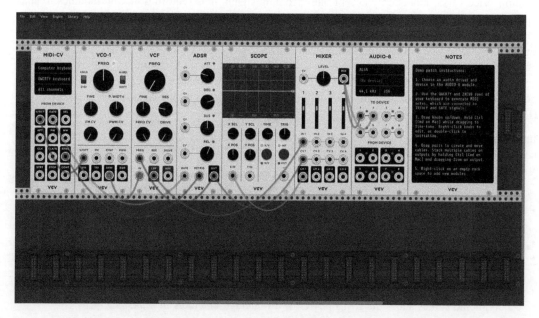

Figure 3.1: The Rack GUI shows modules mounted on the rack rails. When launching the software for the first time, this template patch shows up, implementing a basic monophonic synthesizer.

3.2 Anatomy of a Module

The modules are the functional blocks of your virtual modular synthesizer. They will not do anything useful if you do not connect them. The modules have a limited number of standard components, namely the input and output connectors, knobs and sliders, the switches, and the lights. These components ensure interaction with the module, following a philological approach. Of course, software interfaces offer additional interaction possibilities! For instance, you can design your own graphical elements in C++ and design their reaction to user drag or push. The paid VCV modules have a lot of cool examples (e.g. see VCV Scalar and VCV Parametra) offering a graphical user interface similar to popular DAW plugins. But if you want to keep retro, you can design your own switches, knobs, and lights, as we shall see in later chapters.

The knobs, switches, and sliders are, for the Rack engine, elements that provide input data, and are called parameters. Take note of this term, as we will use it in the chapters devoted to the development of the modules. None of the inputs, outputs, lights, or parameters need necessarily be present. You can even have modules with none of these. Such panels are known as blank panels (more on these in Section 10.8.3).

Interaction with knobs and sliders is done by dragging. Switches can toggle or be pushed by simply clicking with the mouse. Additionally, all parameters (knobs, sliders, and switches) can be reset or modified by right-clicking and inputting a new text value. Double-clicking on a parameter resets to the default value. The exact value of the parameters is also displayed by right-clicking and can be displayed as a tooltip on mouseover if the View → Parameter Tooltips menu item is checked. Cables can be created by dragging from one input to an output, and vice versa. Cables are stackable on outputs, meaning that you can send the same signal to different other inputs, creating a many-to-one relation. Vice versa, inputs only accept one incoming signal, thus requiring mixing modules to sum multiple signals before sending the result to one input. Cables have two modes of operation: monophonic and polyphonic. The first is the obvious one: one cable carries one signal, as in any Eurorack system. The polyphonic mode instead allows you to carry multiple signals into one cable, up to 16. Polyphonic cables are not always available: it depends on the output port they are connected to. You can tell the difference by visual inspection: monophonic cables are thinner than polyphonic ones. Furthermore, the ports will show the signal magnitude with different colors (Figure 3.2).

Indeed, input and output ports show the signal strength and polarity by filling the inner circle with a color according to the following conventions:

- A *positive*, slowly varying mono voltage is shown in *green*, with increasing brightness, reaching the maximum for value > 5 V.
- A *negative*, slowly varying mono voltage is shown in *red*, with increasing brightness, reaching the maximum for a value < −5 V.
- A 0 V signal or a dead input/output has its inner circle filled in *black*.
- Fast oscillations of a mono signal are shown in *yellow*, with increasing brightness with an increased energy content.
- Not a number (NaN) values are shown in *white*.
- Polyphonic cables show the RMS power of all bundled channels in *blue*.

Modules are organized in collections called *plugins*. We should not confuse the terms "plugin" and "module." A plugin, in Rack, is a collection of modules all compiled in one library file and provided

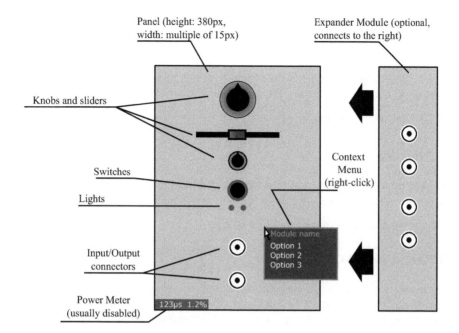

Figure 3.2: Anatomy of a module.

by one author. Of course, an author can develop different plugins, each containing one or more modules.

Plugins may also contain expander modules. These are modules that expand the functionalities of other modules. They may, for example, add ports and knobs to provide additional controls to a parent module. The parent will thus have reduced size, sacrificing some functionalities that can be restored by adding an expander. An expander module must be "connected" to the parent module by placing the two side by side. The expander should go to the right of the parent. This feature mimics expander modules available in the Eurorack world, which are connected by flat cables on the back of the panel.

3.3 Context Menus

Right-clicking on an empty portion of the rack provides the *module browser*. This allows you to search for modules to be added to the rack. The modules are organized by brand or tag and a preview is shown on the side. A quick search can be done by typing some text that filters modules, for example, by name or tag. Once you select the module to be added, this is placed at the position where the right-click was done. If there is not enough space, the new module is placed in the closest empty area.

Right-clicking on a module opens the *module context menu*, with a list of options relevant to that module. By default, all modules have options to:

- initialize a module (i.e. reset its parameters – knobs and switches – to the default state);
- randomize its parameters;
- disconnect cables;

- duplicate a module (including its parameters);
- copy, paste, load, and save a preset (i.e. all parameters);
- disable the module (it will be silent); and
- delete the module.

This menu also shows information about the plugin (website, manual, source code, etc., opening a web browser), and may include factory presets included by the developer.

In Section 8.1.4, you will learn how to add additional items to the context menu.

3.4 Core Modules

Core modules are the only modules provided with Rack by default. They provide audio and MIDI input/output functionalities, a notes module to write down your comments, and a blank panel.

We will briefly review the input/output modules below.

3.4.1 AUDIO-8 and AUDIO-16

The AUDIO module is the interface to the outer world. It allows signals in and out the rack. A sound card and audio driver must be selected. AUDIO-8 provides a maximum of eight inputs and eight outputs, while AUDIO-16 has 16 inputs and 16 outputs. Each pair of inputs or outputs is lit with a green light if the pair is available for the selected sound card. If, for instance, you select a device that only has two inputs and two outputs, only the light for inputs 1 and 2 and outputs 1 and 2 will be lit.

The top part of the panel has three rows, showing the audio driver, the audio device, and its parameters (sample rate and buffer size). Drivers include ALSA and JACK for Linux users, CoreAudio for Mac users, and DirectSound, WASAPI, and any available ASIO driver for Windows users. Once a driver is selected, the second row will show available devices. The last row is split into two fields, allowing you to set any of the supported sample rates and buffer sizes. The larger the buffer size, the higher the latency, but the lower the CPU load and the risk of glitches.

Among the available drivers, you will also find Bridge. If you are using a DAW and you loaded the Bridge VST/AU plugin inside the DAW, you can connect VCV Rack with the DAW by selecting Bridge as the driver. This will get the audio signal to and from the Bridge plugin inside the DAW.

As these are settings available to any DAW or audio software, and are dependent on your hardware and operating system, I will not delve further into the topic. Just bear in mind that any setting that normally works fine with your DAW should be a good starting point for Rack too.

3.4.2 MIDI-CV

This module provides MIDI input to your modular patch. It provides 12 different outputs that translate MIDI functionalities into CV. The module can be monophonic or polyphonic. The top three rows allow you to select a MIDI source by choosing the driver, the device, and the channel. Available drivers include the VCV Bridge, the computer keyboard (yes!), and a gamepad input (yes, yes!). Using the Bridge allows you to get MIDI input from a DAW (e.g. for sequencing events). The computer keyboard allows you to quickly play a patch even if you have no real MIDI device with you. The driver is selected by right-clicking the top row of the module, then you can select the device by right-clicking the second row, and finally the MIDI channel by right-clicking the third row. The available output connectors are:

- *CV.* A V/Oct signal to drive, for example, an oscillator.
- *GATE.* A gate signal with rising edge on MIDI Note On event and falling edge on MIDI Note Off event (GATE). It does not retrigger when playing legato.
- *VEL.* A control voltage proportional to the MIDI Note velocity, ranging 0–10 V, for input MIDI velocity 1–127.
- *AFT.* MIDI Aftertouch.
- *PW.* Reacts to the pitch wheel in the range −5 to +5 V.
- *MW.* Reacts to the modulation wheel in the range −5 to +5 V.
- *CLK.* Supplies pulses timed according to incoming MIDI Clock System Real-Time Messages.
- *CLK/N.* Another source of the MIDI clock but with selectable divider. The divider is selected from the module context menu (CLK/N divider).
- *RTRG.* Generates a 1 ms trigger at any key press (when playing legato, it will generate a pulse even though the gate will not change its state.
- *STRT.* Generates a trigger when receiving a MIDI Start System Real-Time Message.[1]
- *STOP.* Generates a trigger when receiving a MIDI Stop System Real-Time Message.[2]
- *CONT.* Generates a trigger when receiving a MIDI Continue System Real-Time Message.[3]

The module can handle polyphony up to the maximum number of channels, which is 16. Not all outputs can be polyphonic. Modulation and pitch wheels are monophonic due to MIDI limitations. The desired polyphony is set from the context menu. This means that those outputs which allow polyphonic outputs (e.g. gate and velocity) will carry a number of channels greater than one. If one MIDI note is played at a time, only one of the gate channels will be nonzero.

Several polyphony modes are available to manage the routing of the notes into the polyphonic cables channels, and thus into your synth:

- *Reset.* Sends the last received MIDI note message to the first spare output from top to bottom, hence each note occupies an outlet from its onset until the related Note Off message. If all the polyphonic channels are used, the bottom one is overwritten by the new one.
- *Rotate.* Rotates the output connectors in a round-robin fashion from top to bottom, always sending the last received note to the next row of outputs. If more than four notes are active, it does overwrite the next one.
- *Reuse.* Similar to Reset, but if a MIDI note has been previously assigned to one of the rows, it tries to reuse the same row again.
- *MPE.* For MPE controllers.

A Panic option is available in the context menu to reset the module, in cases, for example, of stuck notes.

3.4.3 MIDI-CC

This module translates MIDI Control Change (CC) messages into virtual voltages. Up to 16 CC can be converted into the assignable outputs. Each connector is initially assigned to a CC from 0 to 15, as denoted by the 16-number box on the module panel. To assign different CC to any of the connectors, click the related number (you will see the "LRN" string in place of the number, standing for "learn") and instruct the module by sending a CC message to VCV (e.g. by twisting a knob) or type the CC number on the computer keyboard. Please note that the MIDI learn functionality can work only if you have previously selected the correct MIDI driver, device, and channel from the three upper rows of the module panel.

The outputs send a 0–10 V voltage corresponding to the CC data 0–127. As an exception, some gamepad drivers can generate MIDI values from −128 to 127 that are translated to a voltage from −10 to 10 V.

3.4.4 MIDI-GATE

This utility module is used to send gate signals corresponding to specific notes (e.g. to employ a MIDI controller to trigger events using keys or percussive pads). If connected to MIDI sequencers that send a Note On and Note Off messages in a row, a 1 ms trigger voltage is produced; otherwise, 10 V gate signals are produced in the interval between Note On and Note Off messages.

If you want the gate signal to have amplitude corresponding to the Note On velocity value, you can tick the "Velocity" label in the right-click context menu.

A Panic option is available in this module too, to reset the MIDI status.

3.4.5 MIDI-MAP

This module allows you to map a MIDI-CC message to parameters of any module in the rack. When the mapping is done, a full sweep of the CC from 0 to 127 allows the parameter to go from its minimum to its maximum values. For binary switches, values less than 64 map to 0 and greater than or equal to 64 map to 1.

The procedure to create the mapping follows:

- Click an empty slot – the text changes to "Mapping."
- Click a parameter of a module.
- Send a CC message from your MIDI controller.

A yellow square is printed near the mapped parameter. Maps can be undone by right-clicking them.

3.4.6 CV-MIDI, CV-CC, CV-GATE

These modules translate voltages to MIDI messages, in a similar fashion to their counterparts, MIDI-CV, CC-CV and GATE-CV, respectively.

CV-MIDI translates voltages to MIDI note, aftertouch, and real-time messages. Values are 7-bits-quantized (0–127). Note messages are sent on gate edges or on note change, and are polyphonic. All the other inputs transmit MIDI messages on a change of the quantized value.

CV-CC translates voltage values to selectable MIDI-CC messages, while CV-GATE sends note messages corresponding to the editable values in the module panel when the CV input is over the 1 V threshold.

3.5 Fundamental Modules

These modules are provided by VCV for free and are suggested to anyone who has to gain familiarity with Rack or modular synthesis in general. They cover most basic building blocks of a Eurorack system. A brief description of the currently available modules follows.

VCO-1 is a versatile virtual voltage-controlled oscillator with sine, saw, triangle, and square outputs all available at once on separate outputs. It has exponential frequency modulation with hard/soft sync and pulse width modulation (PWM) for the square wave. It features analog waveform emulation but also digital waveform generation. The analog waveforms feature pitch drift and react to pitch changes with some slew. The digital waveforms have quantized pitch and introduce more aliasing. Figure 3.3 compares the analog and digital sawtooth waveforms.

VCO-2 is a stripped-down version of VCO-1, with morphing between the same waveform type seen in VCO-1.

Fundamental *VCF* is a low-pass/high-pass voltage-controlled filter, emulating a four-pole transistor ladder filter with overdrive and resonance. There is no switch to select the filtering mode, but a low-pass and a high-pass output. The cutoff frequency CV is subject to an integrated *attenuverter* (FREQ CV), while the resonance (RES) and drive (DRIVE) CV are summed to the related knobs. The sum is subject to clipping (i.e. when the RES knob is turned to the maximum level, any positive CV sent to the RES input will not affect the resonance because it already reached the maximum value).

The *ADSR* module generates envelopes to control the evolution of a sound. Its output can be used as a control voltage for any module. It follows the ubiquitous Attack-Decay-Sustain-Release paradigm. Most often it will be used together with the VCA module. The *VCA* module is a voltage-controlled amplifier. In essence, it multiplies the input signal with a control signal in order to shape its amplitude. The response to the control signal can be linear or exponential. If the envelope is sent to the exponential input, the attack and decay ramps will decay "linearly in dB." Confusing, uh? In other words, the decay is exponential, but converted in dB (using the logarithm, which is the inverse operation of the exponentiation) the decay results linear. As shown in Figure 3.4, an exponential decay looks linear on a dB scale, and we can say that it feels more natural and linear to the ear as well.

LFO-1 and *LFO-2* are the low-frequency oscillators in the Fundamental series. Almost similar to voltage-controlled oscillators, they have an extremely low frequency, which goes from tens of seconds per cycle up to 261 Hz (the pitch of C4). LFOs are generally used to modulate a signal and improve expressivity. LFO-2 is more compact than LFO-1 and offers morphing between wave shapes instead of individual outputs. This can be a lot of fun if the wave type is modulated using another signal or LFO.

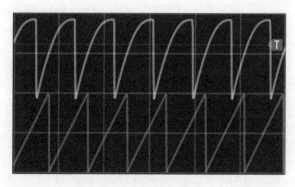

Figure 3.3: Comparison of the analog (top) and digital (bottom) sawtooth waveforms generated by Fundamental VCO-1.

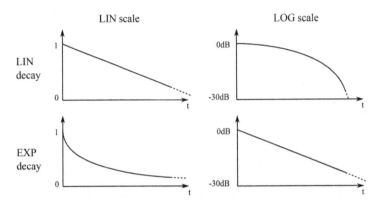

Figure 3.4: Comparison of decay plots. A linear decay decreases by a fixed term per unit of time, looking like a straight line on a linear plot, or as a logarithmic curve on a logarithmic plot (in this case, a dB plot). An exponential decay decreases by a fixed number of decibels per unit of time, thus looking like a line on a dB plot.

Besides the traditional synthesizer building blocks, Fundamental also offers a delay effect, a mixer, and other utilities. Let us examine them.

DELAY is a digital delay with control over feedback, delay time, and tone color. It has a dry/wet knob. The implementation is very good, featuring smoothing on time adjustments, thus avoiding annoying glitches that even hardware effects sometimes have when abrupt changes occur on the delay line length.

A small mixer, *VC MIXER*, is provided. This is a four-channel mixer with independent CV control of each channel level and overall level. It can be used in conjunction with the Core AUDIO module to set levels prior to sending out the signals, but it can be used for many other mixing operation inside your patches.

The attenuverters module, *8VERT*, can also be used to fade signals or even invert their phase. This is a module consisting of eight attenuverters. An attenuverter multiplies the input signal by a gain that ranges from −1 to 1. If the gain is negative, the signal gets inverted (minus sign) and attenuated by the absolute value of the gain. As an extra utility, when no input is connected to a row, the output of that row corresponds to the value of the knob in the range −10 to +10V. This allows the module to create constant CV outputs that can be used to parametrize other modules.

Another utility module, *UNITY,* contains two six-channel direct mixers, with no gain control. It can sum input signals or average them. In the latter case, the inputs are summed and scaled by the number of connected inputs. An inverted phase output is also available (INV).

Fundamental *MUTES* has ten input/output pairs, and a toggle switch for each one, muting the output. Outputs with empty input copy their signal from the first non-empty input above.

SEQ-3 is an eight-step sequencer with three rows of control voltages. It is driven by an internal clock, which has a tempo knob (CLOCK), or by an external signal (EXT CLK). Each time the clock ticks, the sequencer fires a gate signal (GATE output on top) and advances by one step. For each step, a gate signal is also sent individually (bottom row of outputs). For each step, the three control voltages assigned to that step (corresponding to the three rows) are sent to the three related outputs (ROW 1, ROW 2, ROW 3). The clock can be shut off (RUN button) or reset (RESET). The number of steps does not necessarily have to be eight. Using the STEPS knob, a lower number of steps can be used, allowing, for example, the change of the tempo signature. To recap, at each clock step:

- the sequencer advances to the next step;
- the three knob values (one per row) are sent to each of the CV outs (ROW 1, ROW 2, and ROW 3);
- a gate signal is fired to the GATE output; and
- the same gate signal is fired on the GATE OUT output corresponding to the current step (bottom row) – this individual gate signal can be disabled by clicking on the green button close to the output.

A clock generator and a simpler sequencer will be the object of our work in Sections 6.5 and 6.6. If something is not clear enough at this point, it will become clearer later.

Two utility modules, *SS-1* and *SS-2*, are also present for multiplexing and demultiplexing (in short, muxing and demuxing). SS-1 is a demultiplexer that scrolls through the outputs in a round-robin fashion, based on a clock signal. In other words, at each clock pulse, it sends the input signal to the next output. SS-2 is a multiplexer that scrolls through the inputs in a round-robin fashion, based on a clock signal. At each clock pulse, it sends a different input to the output. If this is not totally clear by now, do not worry – more on muxing and demuxing will come in Section 6.3, where we shall build a mux and demux module.

Utility modules to deal with polyphonic cables are *SPLIT*, *MERGE*, *SUM*, and *VIZ*. SPLIT takes a polyphonic cable and splits into monophonic cables. MERGE takes up to 16 input monophonic cables and merges these in one polyphonic output cable. SUM takes a polyphonic input and outputs the sum of all its channels into a monophonic cable. Finally, VIZ takes a polyphonic input and visualizes the intensity of its individual channels through 16 LEDs.

Finally, *SCOPE* is an oscilloscope emulator, with external trigger, internal trigger based on a threshold, X-Y mode, vertical zoom and offset, and time zoom. We shall see in the next section how this works in detail. You should get to master it in order to debug modules.

3.6 Quick Patches

Drawing from the theoretical chapters and our quick introduction to VCV Rack, we now have enough expertise to go through a few examples. These are meant to introduce you to the setup of simple patches, and will help you when you are going to test your modules later.

3.6.1 Audio and MIDI Routing

Ideally, a synthesizer patch should have control inputs and audio outputs. In this case, we consider a MIDI master keyboard as input, sending note messages and control changes. If you like to patch from left to right, I would suggest putting the MIDI inputs on the left and the audio output on the right. For an optimal management of the audio output, some mixing devices are suggested. VCV sells its VCV Console module that does it all very nicely. If you prefer, however, to stay with the free modules, you can stack VCV Mutes and VCV Unity, or VCV Mixer.

In Figure 3.5, an empty rack with MIDI input and audio output is shown. MIDI inputs, placed on the left, include a monophonic MIDI note input and a MIDI-CC input to control the synthesizer's

Figure 3.5: A template patch hosting MIDI inputs and audio outputs.

parameters. For what concerns the audio stuff, we opted for a simple solution that allows four inputs to be routed to a stereo output. The Mixer module has four inputs and faders. The outputs can be mixed with Unity and get routed to the left and right output. For a larger number of inputs, two Mixer modules can be employed and the Mix output of each one can be directly routed to the left and right outputs, respectively. Please note that the AUDIO module inputs and outputs should be understood as the module inputs and outputs, respectively. In other words, you should connect the signal that you want to send to the sound card outputs to the AUDIO module inputs!

By saving such a template and recalling when starting a new patch, you will be able to reduce your working time significantly.

3.6.2 East Coast Synthesis in One Minute

Let us now build a simple East Coast patch starting from the template patch. We want to build a canonical subtractive synthesizer in the East Coast style. This is composed of a cascade of VCO, VCF, and VCA. The VCO generates a rich harmonic tone that is shaped in frequency by a VCF. An envelope generator feeds the VCA to control the note duration and to control its envelope. Finally, a vibrato effect is created by modulating the VCO with an LFO. The modulation wheel control change (CC1 in the MIDI standard) will control the amount of the LFO. This module is monophonic and only has one output that goes to the stereo output pair. Figure 3.6 reports the block diagram for this simple synthesizer.

The patch is shown in Figure 3.7. The MIDI-CV provides GATE, V/OCT, and MW (modulation wheel) outputs. The V/OCT output is sent to the VCO to set the oscillator pitch. The saw output is sent to the VCF for filtering. The low-pass filtered waveform is sent to the VCA to apply an envelope and the VCA output is sent to the mixer. The GATE output triggers an ADSR envelope, used in conjunction with the VCA. We use the first of the two VCAs provided by VCA-2.

The frequency-modulating signal for the vibrato is generated by LFO-1 (sine wave). This wave is multiplied by the MW control voltage using the second VCA module, VCA-2, in linear configuration. The result is sent to VCO-1 as the FM input. The FM CV knob in VCO-1 must be turned to some nonzero value, otherwise in its rest position it kills the modulating signal and no vibrato will take place. The modulating signal is thus the product of the LFO sine with MW and the FM CV knob. When the MW MIDI value is zero, no vibrato will take effect.

Finally, there is a SCOPE to control the output. *Et voila*, this patch emits sounds!

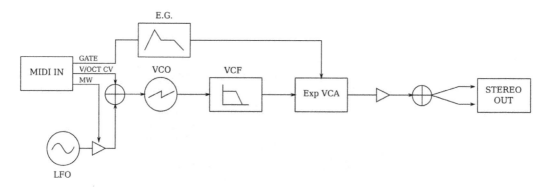

Figure 3.6: A simple East Coast monophonic synthesizer to build with VCV Rack.

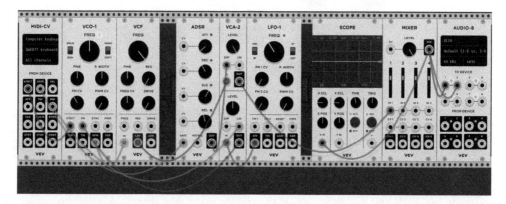

Figure 3.7: A simple East Coast patch with VCO-VCF-VCA cascade, monophonic input, and vibrato.

3.6.3 Using the SCOPE

As any good electronics engineer or practitioner should have experience with the oscilloscope, VCV Rack fans and developers should know how to properly observe signals using an oscilloscope module.

The VCV SCOPE module has all the basic features one needs to master. It has two inputs for signals and an input for an external trigger. The two inputs have independent vertical scaling (zoom) and positioning (for offsetting) knobs. These are X SCL, Y SCL, X POS, and Y POS, respectively, where X and Y are the two inputs. The TIME knob is meant to change the timescale, allowing you to observe high-frequency oscillations and glitches (turning the knob to the right) or longer events such as notes and envelopes.

For operating an oscilloscope fruitfully, it is also important to understand the triggering system. Normally, the signal is sampled at regular intervals, according to the TIME parameter, and stored in a buffer that is printed on the screen. The screen is thus refreshed with a rate that depends on the TIME knob. This mode, which we may call free run mode, is not the only available one. The refresh of the screen can be issued by internal or external triggering. Internal triggering occurs when a rising

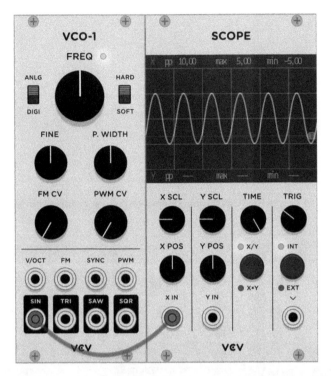

Figure 3.8: Triggering the signal on the first input using the internal trigger.

edge occurs on the X IN and this edge crosses the trigger value. The threshold is selected using the TRIG knob and shown in the SCOPE window as a small arrow indicated with a "T." In Figure 3.9, a sine wave is shown in the SCOPE, starting with a nonzero phase, due to the alignment with a negative trigger. Please note that in absence of X IN, the Scope will not synchronize to the Y IN.

An external trigger can be used as well to synchronize the view with a third signal connected to the EXT input. This resets the view when its value passes the threshold imposed by the TRIG knob.

The SCOPE provides useful statistics related to the X and Y input signals. These are on the top and bottom of the SCOPE screen, written in a tiny font. Peak-to-peak (pp), maximum, and minimum values are provided. By default, one vertical division spans 5 V, and thus the display spans from −10 V to +10 V. If the X SCL or Y SCL knobs are increased or decreased by one step, the vertical division gets halved or doubled, respectively.

Finally, the SCOPE can plot a Lissajous curve in the X-Y view. The X and Y signals control the horizontal and vertical coordinates of the drawing point, respectively. This mode is particularly useful to tune the phase of two signals with same frequency. As you can see in Figure 3.10a, two sines with π phase difference are shown as a sphere. On the contrary, when in perfect phase, they are aligned showing one thin line that goes from bottom left to top right, as shown in Figure 3.10b. In the X-Y configuration, the TIME knob adjusts the persistence of the signal on the screen.

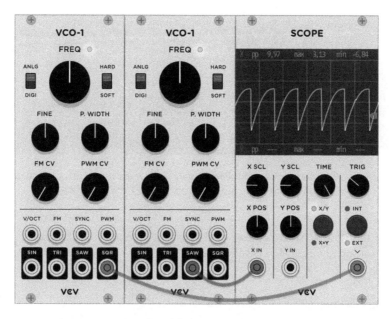

Figure 3.9: Triggering the SCOPE using an external input.

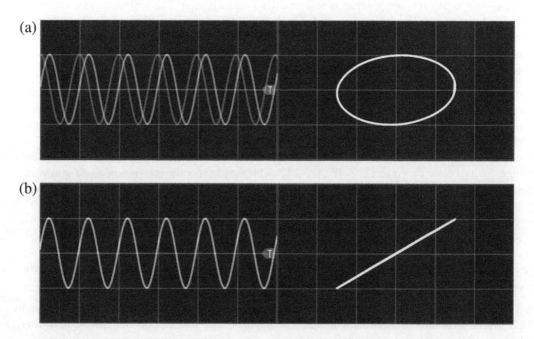

Figure 3.10: The SCOPE in X-Y mode, showing two sine signals with the same frequency, and (a) π phase shift and (b) almost zero phase shift.

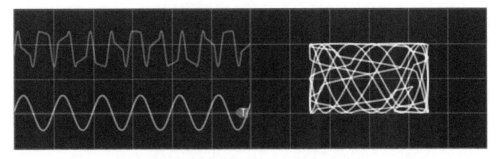

Figure 3.11: Using the SCOPE as a plotter for cool visuals. A sine and a frequency-modulated saw are plotted one against the other (left) and in X-Y mode (right).

Of course, the SCOPE can be used to create cool visuals. In the old-school modular tradition, analog signals were used to drive a cathode-ray tube screen. Figure 3.11 shows the X-Y plot of a frequency-modulated signal.

3.6.4 Observing Aliasing

Sometimes it may be useful to observe signals in the frequency domain. Fundamental does not provide a tool for this, but the ABC collection provided with this book does. This spectrum analyzer looks similar to SCOPE but provides a real-time DFT visualization. The horizontal axis can be linear or logarithmic. The former is useful, for example, to discriminate harmonic partials from inharmonic partials, because the spacing between the harmonics is equal. The logarithmic view is useful for giving the same importance to all frequency bands, following psychoacoustic basics. Specifically, it allows an accurate view of the lower-frequency bands.

We are going to perform a tutorial frequency analysis to watch the presence of aliasing in a signal. Please take note of the fact that the analyzer is very sensitive and even the smallest aliasing components will appear. Figure 3.12 shows the spectrum of a sawtooth tone from VCO-1 (analog mode) at approximately 1 kHz. The aliasing components are visible, although they are hardly noticeable in a listening test because they are tens of dB below the first (correct) 11 partials. In some cases, the aliased partials are not visible, hiding below the correct ones (Figure 3.12). One way to spot the presence of aliasing in such cases is to slightly change the frequency. Sweeping an oscillator is often revealing for the ear as well: you will notice the artifacts given by the inharmonic partials traveling up and down.

3.6.5 Using Polyphonic Cables

In Section 3.6.2, an East Coast monophonic synth was created using Fundamental modules. Here, we are going to make it polyphonic by exploiting polyphonic cables.

Let us first say a few words related to polyphonic cables. This is a feature that is almost unique to VCV Rack. It consists of allowing some ports to handle cables that support multiple signals. It is a little like having a bunch of inputs and outputs in one. A polyphonic cable consists, in other words, of a bundle of up to 16 monophonic cables. When a cable is in polyphonic mode, it is drawn thicker than monophonic cables. But how do you make cables polyphonic?

(a)

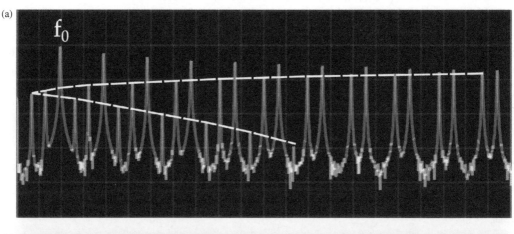

(b)

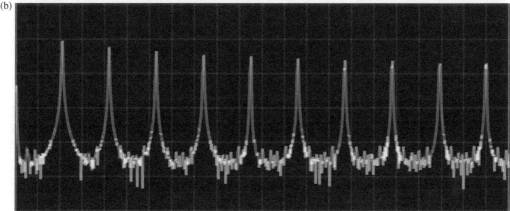

Figure 3.12: The spectrum of a sawtooth wave generated with VCO-1 at approximately 1 kHz. (a) The effect of the aliasing is clearly visible (partials connected with a white line). (b) The exact frequency of the tone is slightly changed so that the aliased partials hide below the skirts of the proper partials. Please note that in the first case, although visible, alias is not noticeable by ear as the undesired partials are tens of dB below the proper ones.

The answer is: it depends on the source module. When connecting ports, the system knows if a cable needs to be poly or mono. If the output port supports more than one channel, the cable automatically becomes polyphonic. The MIDI-CV module, for example, supports mono and poly modes. This setting is available from the context menu, under the Polyphony Channels item. When you drag the V/OCT output of the Core MIDI-CV module with polyphony channels (right-click) set to 1, the cable will be monophonic. If polyphony is set to a number 2–16, the cable will be thicker, indicating that it is a polyphonic cable.

Even though a cable is polyphonic, it can be connected to an input port handling only one channel. What happens in this case? It depends on the module: some will discard all channels but the first one, some will sum all the channels into one. The general rule is:

- Audio inputs should sum all the channels to avoid losing some of them.
- CV inputs or hybrid audio/CV inputs should only take the first one.

More on this will come later. Now let us focus on making the East Coast synthesizer polyphonic. Open the east-coast.vcv patch. Right-click the MIDI-CV module and set any number, from 2 to 16, from the Polyphony Channels menu. Done! Yes, it is simple as that!

The cables stemming from V/OCT and GATE are now polyphonic. As a domino effect, the VCO and the ADSR modules are now polyphonic and their outputs will be polyphonic, thus similarly affecting the VCF, VCA-2, and so on.

Notes

1 When MIDI-CV is connected to VCV Bridge, it receives the Start, Stop, and Continue events from the DAW.
2 Idem.
3 Idem.

CHAPTER 4

Developing with VCV Rack

In Rack parlance, the term "module" refers to the DSP implementation of a virtual Eurorack-style module, while we use the term "plugin" to indicate a collection of modules, all bundled in a folder containing one compiled file, the shared object or dynamic library. Table 4.1 resumes all the terms in the Rack jargon.

VCV Rack modules are similar to the "plugins" or "externals" or "opcodes" used in other popular open-source or commercial standards in that they are compiled into a dynamic library that is loaded at runtime by the application (provided that it passes a first sanity check). However, Rack modules differ in all other regards. We shall highlight some of the differences in Section 4.1, then we shall go straight to showing the components required to create a module (Section 4.2), and then give an overview of basic APIs and classes used to define it and make it work (Section 4.3). At the end of this chapter, we shall start the practical work of setting up your coding environment (Section 4.4) and building your first "Hello world" module (Section 4.5).

Please note that some programming basics are assumed along this part of the book. You should have some experience with C++. We will recall some basic properties of the language at different points of the book as a reminder for users with little experience in C++ or developers who are used to working with other programming languages.

Table 4.1: Some common terms in the Rack jargon

plugin	A collection of Eurorack-style modules implemented in a single dynamic library, developed by a single author or company. The Rack browser lists all plugins and allows you to explore all the modules contained in a plugin. Not to be confused with the term as used in the DAW and VST jargon, where it usually refers to one single virtual instrument.
Plugin	A C++ struct implementing the plugin. We shall use Plugin in monospaced font only when explicitly referring to the C++ struct.
Model	A C++ struct that collects the DSP and the GUI of a Eurorack-style module.
module	Not to be confused with the C++ struct Module. We shall use the generic term "module" to indicate either real or virtual Eurorack-style modules.
Module	A C++ struct that implements the guts of a module (i.e. the DSP part).
ModuleWidget	A C++ struct that implements the graphical aspects of a module.

4.1 Comparison with Other Platforms

Music programming platforms have existed since the 1950s, starting with the first efforts from Max Mathews at Bell Labs USA with MUSIC I. Mathews had a clear vision of how a comprehensive music computing language needed to be designed. For this reason, by the first half of the 1960s, most of the basic concepts were already implemented in his works, with little variations up to these days. Notwithstanding this, in the decades since, there has been a lot of work to design new music computing platforms, to integrate them with new tools, to make them easier to work with, and to update them to the latest programming standards and languages. A lot of these platforms were born in the academic world and are open-source, usually based on C or C++ (think about Pure Data, Csound, SuperCollider, etc.), while some others are commercial (e.g. Steinberg's VST and Apple AU). Others were born in an academic context but later taken as a full-fledged commercial project (e.g. Cycling '74 Max/MSP).

VCV Rack is a commercial project with an open-source codebase, which makes it quite different from the others.

4.1.1 Audio Processing

All the aforementioned music platforms comprise a client application where *audio processing objects* can be loaded and executed. The client application is made of several parallel tasks, managing audio streams, the user interface, handling network connections – for remote audio (Gabrielli and Squartini, 2016) or for licensing and updating – and more. The musical objects usually do the fun stuff. The term "musical object" is just an umbrella term we need here for generalizing all musical programming platforms. In Pure Data, these objects are called *externals*, in Csound *opcodes*, and in VST *plugins*. In Rack, a plugin is rather a collection of objects, and these objects are often called modules, although the proper name for us programmers is *Models*.

All musical platforms are based on the concept of *dynamic linking* of dynamic libraries (i.e. libraries implementing additional functionalities – effects, synthesizers, etc. – that are loaded in the same virtual address space of the client application).

In all the classical music programming platforms, there is an *audio engine* that handles audio I/O and schedules the execution of the audio processing objects in an order that depends on their connections. The audio engine periodically invokes a special function that each audio processing object implements. This function, which for now we will simply call the *audio processing function*, is called periodically for signal processing to take place. The audio processing function feeds the object with new audio data and expects processed audio data to be returned. Of course, this is not mandatory. Some objects only take input data (e.g. a VU meter) and do not process it. Other objects have no audio input, but return audio data (e.g. function generators).

The audio engine period is determined by the data rate (i.e. $P = B/F_s$), where F_s is the audio sample rate and B is the number of samples to process at each call to the processing function (i.e. the size of the buffer to be processed).

Suppose that we have a sample rate of 48 kHz and we process 64 samples at each iteration. The maximum execution time E_{64} available is $E_{64} < P = \dfrac{64}{48,000} = 1.\bar{3}\ ms$, where the "minor than" is to stress the fact that we need to get below the theoretical limit P. In this time frame, the computer needs to not only execute our code, but also all other operating system tasks.

We can also define the *real-time factor* as the ratio between an execution time and the engine period:

$$RT = \frac{E}{P} \ [\%] \tag{4.1}$$

usually exposed as a percentage. It is obvious that RT must always be less than 100%.
Please note that the audio engine requires some time to pass the output to the sound card, and since we are working with general-purpose computers there is always some random interrupt incoming that has priority over our audio application, so a stable audio processing without glitches can be obtained only with RT factors well below 100% (this is very sensitive on the operating system, the operating system scheduler, the audio driver and many other factors).

For most musical programming applications, the buffer size is a power of 2 and can often be selected by the user to obtain a trade-off between latency and computational resources. Pure Data, for example, by default has a buffer size of 64 samples. To that extent, however, Rack differs from all the other platforms: *only one sample at a time is passed to the periodical processing function*, called **process(...)**.

The definition of RT factor for Rack changes to:

$$RT = \frac{E}{T} \ [\%] \tag{4.2}$$

because in this case $B = 1$. The maximum execution time is now $E_1 < \dfrac{1}{48,000} = 20.8 \ \mu s$, lower than the case above. Fortunately, in this short time, we just have to compute one sample, not 64. If the execution would be linearly proportional to the number of samples to process, then the following would hold: $E_{64} = 64 \cdot E_1$. In that case, the RT factor would stay the same for batch processing and for sample-wise processing. However, in practice, it turns out that processing one sample at a time will reduce chances to optimize code and exploit instruction pipelining and parallelization. Furthermore, a lot of overhead is added. It thus turns out that $E_{64} < 64 \cdot E_1$.

If you are wondering why sample-wise processing is less efficient, think about a factory with an assembly line, processing a batch of 100 shoes compared to a single craftsman producing one shoe at a time. The assembly line has specialized workers, each excelling at an operation. Shoes are moved in batch from one machine to another. The assembly line will take less time to make each shoe than the craftsman does, making the factory more efficient (although questions of quality and ethics arise).

At this point, you may be wondering why, in VCV Rack, we are not exploiting the efficiency inherent to process large buffers. In Rack, the main objective is the simulation of electronic circuits, with the ability to create near delayless loops, allowing signal feedback like in modular synthesizers.

Imagine the cascade of a module "A" and a second module "B." Let the output of "B" be fed back to "A." In an analog environment, we have a delayless loop. Instantaneous hardware feedback is a key concept for so-called *No-Input Mixing* music performances. In a discrete-time setup, however, the feedback is delayed by – at least – one sample. Suppose that processing is sample-wise: the computer must, first, execute the instructions for module "A" and extract one output sample. The output is used by module "B" to compute its output sample. At the next time instant, the output of "B" is fed back to "A" and the cycle restarts. If, however, modules A and B are programmed to compute N output values from N input values, the feedback delay increases to N: samples will be computed in batches of N and then fed to the next module. With increasing N, the simulation of a feedback electronic system departs from reality. The only way to get close to a real analog system is thus to process the smallest unit of signal (i.e. a single sample) and make it as short as possible (i.e. increasing the sample rate). The higher the sample rate, the closer to the hardware system, because the time step gets closer to zero.

Sample-wise processing is a necessary ingredient to make Rack get very close to the behavior of modular systems. Of course, this comes with some CPU sacrifice.

4.1.2 Scheduling

Since there is a *scheduling* order for the audio processing objects, parallel execution of objects in many cases is not an option. Imagine a chain of three objects, implementing an oscillator, a filter, and an amplifier. The audio engine will schedule them serially, as the second needs to wait for the execution of the first, and the third needs the second one to be executed. The audio engine lets each one of them run and return the processed signal, which is subsequently fed to the next object. After the third object has finished, the audio engine can pass the processed signal on to the sound card. This implies that any of the objects may potentially be stealing CPU time to the others. If the available execution time is expired while they have not yet finished processing, there will be audio glitches, as the sound card has no new data to convert and will recycle old data or fill with zeros. The RT factor must be lower than 100% to avoid audio dropouts (i.e. loss of data).

As we said, all music programming platforms have an audio processing function. They have other periodically called functions too. Control data may come in a synchronous or asynchronous way. In the first case, a rate will apply, although lower than the audio engine rate. In the second case, asynchronous events may arise, such as MIDI events or user events. In this context, all platforms exhibit differences. Max, for example, processes control data at a fixed rate of 1 kHz. *Rack does not have any control data.* This is one of its prominent characteristics: as in modular synthesizers, all information is transmitted by a (virtual) voltage.

Fortunately, current CPUs have multiple cores. Rack is thus able to distribute computing tasks among one or more cores. This means that several process routines are computed at the same time, reducing the risk of dropouts.

Other periodic functions are the GUI rendering functions. In Rack, each graphical widget has a draw method that is called periodically at a rate, ideally, of 60 frames per second.

From this discussion, it should be clear that any musical platform enforces the developer to strip all tasks that are not time-critical or audio-related from the audio processing function, to make it quick. All these accessory tasks should be handled separately in other processes that may be periodically or asynchronously called (Figure 4.1).

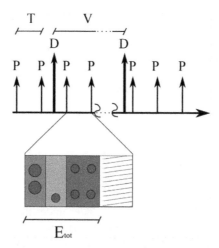

Figure 4.1: Scheduling of the process functions "P" and the draw functions "D" in VCV Rack in a single-core scenario. The process functions are invoked with a period equal to the sample time (i.e. $T = 1/F_s$), where F_s is the audio sampling rate (e.g. 44,100 Hz). The draw functions are called with a larger period of $V = 1/F_v$, with F_v being the video rate, usually 60 Hz. During the sample time, all the modules must perform their process function. The sum of all their execution times E_{tot} is the total execution time of the patch. The spare time is left to all other lower-priority tasks. The real-time factor RT is thus E_{tot}/T.

4.1.3 Polyphonic Cables

Polyphonic cables are one of the big differences with other modular environments. Although the concept is quite simple, this feature is not available in other common platforms, so their use may not be straightforward at the beginning. Polyphonic cables bundle up to 16 signals into one wire. In essence, all cables in Rack are polyphonic, however, when they revert to monophonic mode if the output port they are connected to is not poly-capable. In such cases, they are shown as thinner. Vice versa, when a polyphonic cable is connected to a mono-only capable input, this may be programmed to discard all channels but the first[1] or to sum all of the input components in one.

4.1.4 Expander Modules

Another interesting feature for developers is the possibility to create expander modules. These are modules that do not live on their own, but can be attached to a compatible parent module to expand their functionalities and I/O capabilities. From the developer point of view, expander modules are handled using pointers, allowing any module to be chained with other ones that are on the immediate left and right of it. This feature is not available in platforms such as Pure Data, Max, VST, and so on.

4.1.5 Simplicity and Object-Oriented Programming

One last feature that is unique to Rack is its simple and neat API. This is partly due to the fact that its scope is quite constrained, and a lot of stuff that other platforms have has been wiped away from Rack. Its simplicity is also due to the fact that the API is written from scratch, with the high vantage point of decades of computer music API and with the additional advantage of adopting the latest OOP and C++ standards. Consider the fact that older platforms are starting to wrap their codebase to allow C++ code on top of a bare C core (e.g. this is the case of Csound (Lazzarini, 2017)).

Finally, the ability to write sample-wise processing code makes developing much easier and makes the code easier to read. A lot of aspiring developers started writing code for Rack just by learning from the first few examples provided by VCV (Fundamental, Befaco, and Audible Instruments plugins).

After all, if Rack was not so developer-friendly, this book would never have even been thought of. Credit for all this goes to VCV Rack's main developer.

4.2 Plugins: Files and Directories

A Rack plugin, practically speaking, requires several components to be found by the application at startup.[2] Plugins can be located either in the plugins folder under the installation directory or in the plugins folder under the *local Rack directory*. The local Rack directory is different for each operating system: it is "My Documents/Rack/" on Windows, "Documents/Rack/" on Mac and "~/.Rack/" on Linux. Each plugin will have a directory with corresponding name under one of the aforementioned plugins folders. The components that are found under each plugin folder are:

- The compiled object, with different extension depending on the operating system: plugin.so (Linux), plugin.dll (Windows), or plugin.dylib (macOS). This is also known as "shared object" in Linux parlance or "dynamic-link library" under Windows or macOS;
- The res/folder, usually including SVG graphic files (background) and additional resources (fonts, other graphical elements, etc.);
- Optional but important: licensing files and a readme.

The plugins are searched by VCV Rack during startup by looking at all folders under Rack/plugins/ and under the local Rack directory. For each folder, the compiled object (plugin.dll or .so or .dylib) file is opened and verified, and if the plugin passes a sanity check it can be loaded.

In the development of plugins, you will also need the source code. Generally, you have a *makefile* script to compile the source code, under the plugins root (i.e. plugins/<myplugin>/) and the C++ source code under the src/ folder (plugins/<yourplugin>/src/). A typical arrangement of the folder tree is provided in Figure 4.2.

The compiled object is the core of your project, and results from compiling all your .cpp and .hpp files. The compile process scans the *include* files for functions, classes, and variables you recall from your code, such as standard C library function (printf, memset, etc.) and specific components such as the functions and classes from Rack. Since the development of the Rack codebase is very quick, the components you include from Rack in your code will vary from version to version. It is thus important that the compiler can find out the correct version of the Rack APIs, the one you mean to use from your code. During compilation, it is also important that the linker can find compiled elements (shared objects and libraries) that match the included files.

Similarly, if you compile a plugin by including and linking files pertaining to a specific version of Rack, you cannot open the same plugin with a different version of Rack. When Rack scans your plugin folder, it will try to load the shared object it finds in it; however, it will give up if there is some mismatch. An example of the warning it will give when a plugin cannot be loaded follows:

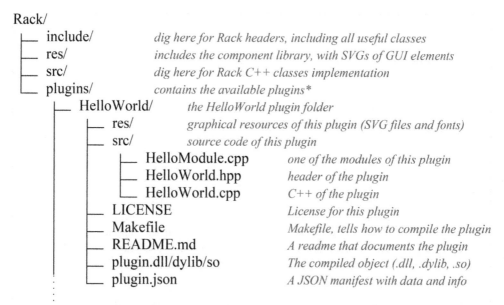

Figure 4.2: **The directory tree of Rack and its plugins.**

```
[warning] Failed to load
library ./plugins/ABC/plugin.so: ./plugins/ABC/plugin.so:
undefined symbol: <symbol-name-here>
```

For this reason, it is crucial to compile and run for the right version of Rack. The "right" version is the one that provides not only the same API, but also the same ABI (application binary interface), that was used for compiling Rack, and thus symbols match.

4.3 Plugins: Software Components

As stated before, a Rack plugin is a collection of *models*. Each model implements a module and a widget. In a few words, the module describes the inner behavior of the model in relation to voltages and parameters, while the second one describes the graphical appearance of the model and its interaction with the module. Dividing the tasks between module and widget also allows you to create different GUIs for the same module, preview a GUI by creating a model without a module, or run DSP code in a model that has a module but no widget.

In terms of code, Rack provides two main objects:

*   the `Module` struct; and
*   the `ModuleWidget` struct.

We will inherit these two objects to define our models. The `Module` has fundamental members such as the `process(...)` function, which will be executed at each audio engine cycle, while `ModuleWidget` instantiates all graphical elements and any other object you may need.

`Module` is a struct, containing the following important methods and members:

```
std::vector<Param> params;
std::vector<Input> inputs;
std::vector<Output> outputs;
std::vector<Light> lights;

virtual void process(const ProcessArgs &args)
virtual void onSampleRateChange()
void config(int numParams, int numInputs, int numOutputs, int
numLights = 0);
void configParam(int paramId, float minValue, float maxValue, float
defaultValue, std::string label = "", std::string unit = "", float
displayBase = 0.f, float displayMultiplier = 1.f, float displayOffset = 0.f)
```

The first four members are vectors of parameters (e.g. knobs), input and output connectors, and lights that are initialized in the constructor. This means that when we subclass the Module struct, we are able to impose a certain number of these elements. This is hard-coded (we cannot add an output during execution, but this makes absolute sense for a virtual modular system).

The next four methods are important for our scopes. *The process method is the fundamental method for signal processing!* It is called periodically, once per sample, as discussed in Section 4.1. It takes a constant argument as input: the ProcessArgs structure, that contains global information such as the sampling rate. The config method is necessary to indicate the number of parameters, ports, and lights to create. The configParam describes the range of a parameter, its labels, and more.

Other useful methods that can be overridden are onSampleRateChange, to handle special conditions at the change of the internal engine sample rate, onReset, to handle special conditions during reinitialization of the module, and onRandomize, to make special randomization of the parameters.

The ModuleWidget is the object related to the appearance of the module, and it hosts all GUI elements such as knobs, ports, and so on. It also handles mouse events such as mouse clicks that can be managed by developing custom code:

```
Module *module = NULL;
std::vector<ParamWidget*> params;
std::vector<PortWidget*> outputs;
std::vector<PortWidget*> inputs;

void draw(const DrawArgs &args) override;
void onButton(const event::Button &e) override;
void setPanel(std::shared_ptr<Svg> svg);
void addParam(ParamWidget *param);
void addOutput(PortWidget *output);
void addInput(PortWidget *input);
virtual void appendContextMenu(ui::Menu *menu) {}
```

The development of novel models is made easy by Rack APIs. Among these, we want to highlight basic components such as:

- a library of GUI elements, from abstract widgets to knobs and switches (examined in Section 5.2);
- DSP and signal processing functions and objects, from the Schmitt Trigger (in include/dsp/ digital.hpp) to interpolation (in include/math.hpp), from efficient convolution (include/dsp/fir. hpp) to fast Fourier transform (in include/dsp/fft.hpp); and
- various utilities such as random generator functions, string handling (in include/string.hpp), or clamping and complex multiplications (in include/math.hpp).

We will explore more along these pages; however, you are encouraged to dive into the source code of Rack to inspect all the pre-implemented utilities it features.

4.4 Setting Up Your System for Coding

In this section, we see how to setup a system for building Rack.

The procedure differs for all three operating systems supported, Linux, Windows, and macOS. In all three cases, you will need to work with a terminal (aka console) using Unix-style commands.

For developers who are not used to a Unix terminal, we report a shortlist of bash commands that are useful when working on the terminal to handle your files and folders. The macOS and Linux terminal commands are almost the same. On Windows, you need to install a mingw shell that behaves the same. Some basic commands are:

- `cd <dirname>` change to directory <dirname> (remember that the parent directory is denoted as ../)
- `mkdir <dirname>` creates a new directory <dirname>
- `rm <filename>` removes a file
- `rm -rf <dirname>` removes a directory recursively
- `cp <file> <dest>` copy a file to destination folder or folder/name
- `cp -r <folder> <dest>` copy a folder recursively to a destination path
- `mv <dest>` moves a file or folder src to a destination path or name (you can use it both to rename or to move)

In the following subsections, you will find quick tips to set up your system. However, considering the fast-paced evolution of Rack, things may change slightly in future releases. Check the online Rack manual for changes (https://vcvrack.com/manual/Building.html).

4.4.1 Linux

Several packages need to be installed from your package manager, if not already present.

On Ubuntu, for example, open a terminal window (Ctrl+Alt+T) and type:

```
sudo apt install git gdb curl cmake libx11-dev libglu1-mesa-dev
libxrandr-dev libxinerama-dev libxcursor-dev libxi-dev zlib1g-dev
libasound2-dev libgtk2.0-dev libjack-jackd2-dev jq
```

On Arch Linux:

```
pacman -S git wget gcc gdb make cmake tar unzip zip curl jq
```

That's it.

4.4.2 Windows

You need to install MSYS2, a software package that allows you to manage the building process as in Unix-like platforms. It also features a package manager, very useful to get the right tools easily with one console command. Currently, MSYS2 is hosted at www.msys2.org/, where you can also find the 32- or 64-bit installers. You need to install the 64-bit version. Please also note that Rack is not supported on 32-bit OSs, so it would make no sense anyway to install 32-bit tools.

Once done, launch the 64-bit shell (open the Start menu and look for mingw64 shell) and issue the command:

```
pacman -Syu
```

Then restart the shell and type:

```
pacman -Su git wget make tar unzip zip mingw-w64-x86_64-gcc mingw-
w64-x86_64-gdb mingw-w64-x86_64-cmake autoconf automake mingw-w64-
x86_64-libtool mingw-w64-x86_64-jq
```

This will tell the MSYS2 package manager to install the required software tools.

4.4.3 macOS

You need to install Xcode with your Apple ID. Then install Homebrew, a command-line package manager that will help you install the required tools. Go to https://brew.sh/ and follow the instructions to install Homebrew (it is generally done by pasting a command into a terminal window). After Homebrew is ready, in the same terminal you can type:

```
brew install git wget cmake autoconf automake libtool jq
```

This will get you started.

4.4.4 Building VCV Rack from Sources

Once your system is ready you can try building Rack from sources. This procedure is the same for all operating systems. You can decide to compile Rack from the latest sources or get a stable version from the GitHub repository that hosts its code. I suggest you work on an easy-to-reach folder and create a path where you will be able to hold several versions of Rack all compiled, in order to port, backport, or debug your code. On Linux, I use a path such as:

```
/home/myusername/Apps/Rack/
```

and on Windows I use the following path:

C:\Rack\   (*Note*: Using MSYS2 shell, you can cd there using: cd /c/Rack)

Inside the Rack folder, I have different versions of Rack (e.g. Rack050, Rack051, Rack-latest, etc.). Start with:

```
git clone https://github.com/VCVRack/Rack.git
```

This will prepare a local Git repository and download its latest content. If you want instead to go straight with one specific branch (e.g. branch "v1"), you can do the following:

```
git clone --single-branch --branch v1 https://github.com/VCVRack/
Rack.git
```

Step in the freshly created repository folder:

```
cd Rack
```

Now you need to download other software modules required by Rack available from Git:

```
git submodule update --init --recursive
```

and other software dependencies that are automatically downloaded and prepared, invoking the following command:

```
make dep
```

This step will take some time and require that you have an Internet connection active in order to download packages. It may happen that the required packages cannot be downloaded for temporary server errors. Take a look at the output. All operations need to be successful, otherwise you will miss fundamental software components and will not be able to compile and run.

The real compilation of Rack is issued by:

```
make
```

And finally, drums rolling:

```
make run
```

will start your compiled version of Rack. Please note that make run will start Rack in development mode, disabling the plugin manager and loading plugins from the local plugins folder.

If you want to speed up the compilation process, you can issue all the make commands adding the "-j <parallel jobs>" option. This starts several parallel jobs, up to the number of parallel threads your CPU supports.

You may encounter some issues in the process. It is very important that you collect the output of the terminal before asking for suggestions on the official forum, on the GitHub issue tracker, or to users of the Rack developers Facebook page.

4.5 Building Your First "Hello World" Module

To build a plugin, you first need to have a working Rack environment compiled from source. The precompiled Rack application obtained from the official VCV web page does not allow developing, so you need to get and prepare a development version alongside it. Go back to Section 4.4 for the details.

The online resources of this book provide a simple "Hello World" module to verify that your system is ready to compile and work with developed modules. Download the HelloWorld plugin folder to the directory of installation of Rack under Rack/plugins/. You will find the source files under src/ and the graphical resources under res/. This plugin contains one module showing a blank screen with the text "Hello World," following the tradition of coding textbooks.

4.5.1 Build and Run

We first check whether the whole plugin builds fine. Open the terminal, go to the plugin folder, and build using the makefile:

```
cd Rack/plugins/HelloWorld

make
```

This should be sufficient, and you should see one of the following files in the HelloWorld root, depending on the operating system:

- plugin.so (Linux)
- plugin.dll (Windows)
- plugin.dylib (macOS)

Run Rack to verify that the HelloWorld module is loaded (Figure 4.3):

```
cd ../../

make run
```

Fine, but we can do better than this! However, before starting programming, we should take a look at the source code skeleton we have built so far.

4.5.2 Structure of the Source Files

This section reports the structure of the source files. Keep this section bookmarked whenever you need to create a new plugin or you don't recall the basic skeleton of a module.

The elements involved with this plugin are:

- *plugin.json.* A JSON manifest (i.e. a file containing the information related to the plugin and its modules with a syntax readable to both machines and humans).
- *src/HelloModule.cpp.* Implements the module we see and its guts.
- *src/HelloWorld.hpp and src/HelloWorld.cpp.* Defines and initializes some pointers and data that allows the plugin (i.e. the module collection) to work.

Figure 4.3: The Hello World module.

- *res/CTemplate.svg.* The background image file for the module. We shall use this throughout the whole book to give our modules a simple yet elegant look and make them look similar.
- *Makefile.* This is the project makefile (i.e. what is executed after the make command is issued). You don't need to supply this to other users.
- *LICENSE.* You should always add a license file to your modules.
- *build/.* This folder is created automatically for building. You can disregard its content, and you don't need to supply this to other users.
- The plugin dynamic library (plugin.so, plugin.dll, or plugin.dylib, depending on the OS).

The plugin.json file contains data regarding the plugin, including name, authorship and licence, and the modules included in the plugin, with name description and tags. Allowed tags are listed in Rack/src/plugin.cpp (see `const std::set<std::string> allowedTags`). The plugin.json for the HelloWorld plugin follows:

```
{
  "slug": "HelloWorld",
  "name": "Hello World Example",
  "brand": "LOGinstruments",
  "version": "1.0.0",
```

```
"license": "CC0-1.0",
"author": "L.Gabrielli",
"authorEmail": "l.gabrielli@univpm.it",
"authorUrl": "www.leonardo-gabrielli.info/vcv-book",
"sourceUrl": "www.leonardo-gabrielli.info/vcv-book",
"modules": [
        {
            "slug": "HelloWorld",
            "name": "HellowWorld",
            "description": "Empty Module for Demonstration Purpose",
            "tags": ["Blank"]
        }
    ]
}
```

Note: The plugin version number must follow a precise rule. From Rack 1.0 onward, it is R_MAJOR. P_MAJOR. P_MINOR, where R_MAJOR is the major version of Rack for which you provide compatibility (e.g. 1 for Rack 1.0) and P_MAJOR and P_MINOR correspond to the version of your plugin (e.g. 2.4 if you provided four minor changes to the second major revision of your plugin).

The makefile is a script that instructs the make utility how to build your plugin. You will notice that it includes ../../plugin.mk, which in turn includes ../../arch.mk, a section to spot the architecture of your PC, and ../../compile.mk, the part containing all the compilation flags.

The source code of the plugin includes at least a header and a C++ file, which take the name from the plugin (e.g. HelloWorld.hpp and HelloWorld.cpp). The former gives access to the Rack API by including rack.hpp and declares a pointer to the plugin type. It also declares external *models* (i.e. the meta-objects) that contain the ModuleWidget (i.e. the GUI) and the Module (i.e. the DSP code) for each one of your virtual modules. The HelloWorld.hpp file looks as follows:

```
#include "rack.hpp"

using namespace rack;

extern Plugin *pluginInstance;

extern Model *modelHello;
```

In HelloWorld.cpp, we instantiate the plugin pointer and we initialize it, adding our modules to it with the addModel() method:

```
#include "HelloWorld.hpp"

Plugin *pluginInstance;

void init(Plugin *p) {
    pluginInstance = p;
```

```
    p->addModel(modelHello);

}
```

Each model is implemented in a separate C++ file. Any such file will include:

- The `Module` child class, defining the `process(...)` method and any other method related to the signal processing.
- The `ModuleWidget` child class, defining the GUI and other properties of the module, such as the context menu, and so on.
- The dynamic allocation of the `Model` pointer.

The `Module` is subclassed as follows:

```cpp
struct HelloModule: Module {
    enum ParamIds {
        NUM_PARAMS,
    };

    enum InputIds {
        NUM_INPUTS,
    };

    enum OutputIds {
        NUM_OUTPUTS,
    };

    enum LightsIds {
        NUM_LIGHTS,
    };

HelloModule() {
    config(NUM_PARAMS, NUM_INPUTS, NUM_OUTPUTS, NUM_LIGHTS);
}

    void process(const ProcessArgs &args) override {

        ;

    }

};
```

As you can see, objects are implemented as a struct. In C++, `class` and `struct` are almost equivalent, differing only in the fact that a `struct` defaults all methods and members to public, while for a `class` they default to private. For this reason, throughout the book we will sometimes use the term "class" as a synonym of "object," generically including both "struct" and "class" under this broad term, when both are to be addressed. Since we are following an OOP paradigm, it is worth stressing the difference with a C struct that is intended only as a collection of variables (although compilers nowadays allow struct to have methods even in a C project).

The enums are empty, as you see, and the compiler will default NUM_PARAMS, NUM_INPUTS, NUM_OUTPUTS, and NUM_LIGHTS to zero. There is no need to use the enums; we could just say:

config(0, 0, 0, 0);

However, you'd better keep this skeleton module more general, to have a little more flexibility in all the practical cases, when you will need to add inputs, outputs, lights, or knobs. Most of the modules we are going to create have non-zero NUM_* enums.

As you can see, the class also declares the process(...) method, stating that it will override the same method from the base class Module. The implementation is empty for now as the HelloModule class does not define any action on signals; it is just a blank panel (more on blank panels in Section 10.8.3).

We then define the module widget child class and implement its constructor:

```
/* MODULE WIDGET */
struct HelloModuleWidget: ModuleWidget {
        HelloModuleWidget(HelloModule* module) {

                setModule(module);
                setPanel(APP->window->loadSvg(asset::plugin
(pluginInstance, "res/HelloModule.svg")));

        }

};
```

As you can see, HelloModuleWidget inherits the ModuleWidget class. The constructor needs to have a pointer to the Module subclass it will handle. More on this later. The setPanel method takes an SVG file, loaded with loadSvg, and applies it as background to the module panel. *Please remember that, by default, the module size will be adapted to the size of the SVG file.* The SVG file must thus respect the size specifications, which impose the height of the module to be 128.5 mm (3 HU in the Eurorack format) and the width to be an integer multiple of 5.08 mm (1 HP). We will discuss the creation of the SVG panel in more detail in Chapter 5. Please note that the panel size can be changed from code, as we shall see for the ABC plugins.

When the zoom is 100%, a module height of 128.5 mm corresponds to 380 pixels:

#define RACK_GRID_HEIGHT 380 // include/app/common.hpp

Finally, the file ends with the creation of the Model, indicating the Module and the ModuleWidget as template arguments:

Model * modelHello = createModel<HelloModule, HelloModuleWidget>("HelloWorld");

The input argument takes the slug of the module. The template arguments are the subclass of Module and ModuleWidget.

Now Rack has all the required bits to compile your Model and add it to the plugin.

4.6 Adopting an IDE for Development: Eclipse

Now you have the *basic* tools for compiling a project. However, writing and managing code can be complex sometimes. You will need a text editor with syntax highlighting at least. Each operating system has its own good editors. If you additionally want to manage your source code, watch your

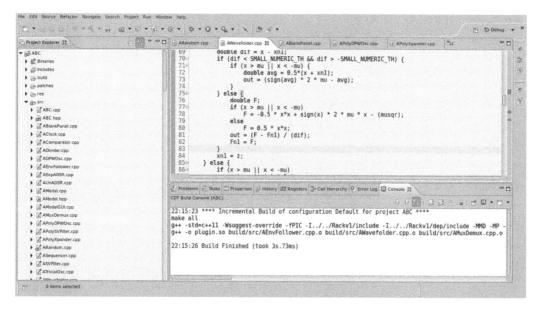

Figure 4.4: The Eclipse IDE, with the Project Explorer (left panel) for browsing projects and files, the main panel (top right) showing multiple source files, and the Console panel (bottom right) showing the makefile log after building the ABC project.

project structure, launch a build, and have automatic suggestions and code verification, then an integrated development environment (IDE) is better suited.

One option I would suggest to all readers is the Eclipse platform,[3] maintained by the Eclipse foundation, open-source and highly flexible. This tool is cross-platform (Linux, Windows, macOS) and is available in different flavors, depending on needs and the language you use. For C/C++ developing, I recommend downloading a build of the C/C++ developers IDE, or adding the Eclipse CDT package to an existing Eclipse installation.

Eclipse (and similarly most C/C++ IDEs) shows up with several distinct panels, for efficient project management. The Eclipse CDT shows a main panel with multiple tabs for reading and writing the source code, a Project Explorer for browsing projects and the related files, and additional panels for searching keywords, building, debugging, and so on. A typical Eclipse window is shown in Figure 4.4.

4.6.1 Importing a "Makefile Project"

Any existing project with a makefile can be imported, and to exploit automatic C/C++ code indexing the Rack source can be referenced. We shall describe these few steps here, leaving the reader to the online Eclipse documentation or other reference texts for instructions to use Eclipse.

We will *first import the Rack codebase, needed by the Eclipse C/C++ indexing tool*. The indexing tool is what makes automatic suggestion and automatic error/warning detection possible. Open Eclipse and go to File → New → Makefile Project with Existing Code. Select a location and give the project a name. The location should be the root of your Rack installation (i.e. where the makefile

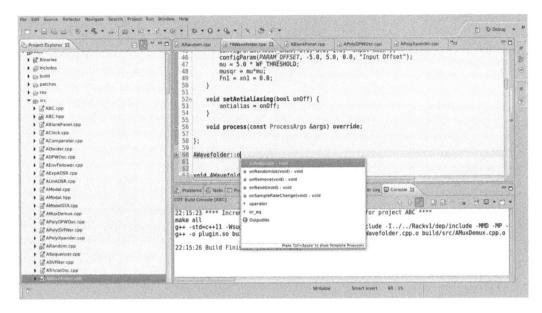

Figure 4.5: Eclipse code editor showing automatic suggestions, available by pressing Ctrl+Space.

resides). I suggest including the version in the project name (e.g. Rack101 for v1.0.1), so you will be able to add future versions to Rack without having to delete previous ones. The toolchain is not important – you can leave it to <None>.

Now, similarly, we will import the ABC library (or any other plugin folder) with File → New → Makefile Project with Existing Code. The project location should be the ABC folder, where the makefile resides, and I suggest you include the version of Rack you are compiling against in the project name.

Now you have the ABC project in the project explorer. Right-click it and go to Properties. Select *Project References* and check the Rack project name. Now the indexer will include Rack in its search for classes and declarations. To force building the C/C++ index, go to Project → C/C++ Index → Rebuild.

Now open a source file (e.g. AComparator). If you *Ctrl+Click* a Rack object name (e.g. Module), *Eclipse will take you to its declaration.* This speeds up your learning curve and development time!

You can also experience automatic suggestions. Automatic suggestions are available by pressing Ctrl+Space. The Eclipse indexing tool will scan for available suggestions. By typing in more letters, you will refine the research by limiting it to words starting with the typed letters. If you write, for example, the letter "A" and then press Ctrl+Space, you will get a large number of suggested objects, functions, and so on. However, if you type "AEx," you will get only "AExpADSR," a class that we have implemented in our ABC project (Figure 4.5).

4.6.2 Building the Project Inside Eclipse

Any C/C++ project in Eclipse can be built from the IDE itself. If we import the project as a "Makefile Project," the IDE will know that it just has to issue the make command as we would do ourselves into

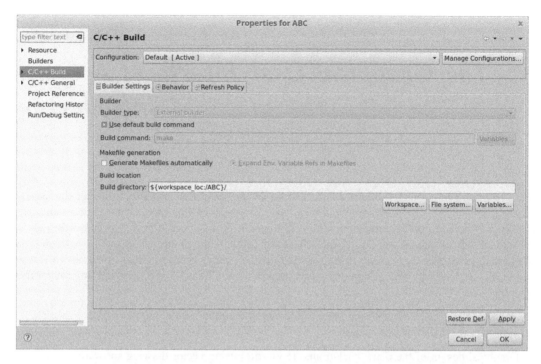

Figure 4.6: The Project settings window, showing default settings to be used for building the ABC project. As you can see, the build command is the default one (make).

a terminal. You can check this by right-clicking on the name of the project in the Project Explorer and clicking Properties, or going to Project → Properties. On the left, you will see C/C++ Build, and the Builder Settings should be set by default so that it uses the default build command ("make"). The build directory is the root folder of your plugin. This is exemplified in Figure 4.6.

To test whether the Build command works fine, go to Project → Build Project. You should see the log of the build process in the Console window, as seen in Figure 4.4. You should see no errors in the log.

This section was a small hint to setting up your Eclipse IDE and getting ready to go. We will not go further into the vast topic of developing with Eclipse, as it would be out of our scope, but you are encouraged to read the documentation, or compare with any other IDE you are already familiar with.

Notes

1 This is the suggested behavior that should be followed by all developers.
2 Rack loads the plugins only once at startup, so you need to restart Rack if any of the plugins have been updated.
3 Experienced developers will have their own choice for the IDE, and some old-school developers will prefer developing from a terminal using applications such as Vi or Emacs. Some will complain that I'm suggesting this IDE instead of another one. That's fine – let's make peace, not war. Eclipse will suit all platforms and integrates nicely with VCV Rack – that's why I'm suggesting it in the first place.

CHAPTER 5

The Graphical User Interface: A Quick Introduction

We all recognize how much the look and feel of hardware synthesizer modules influence the user at first sight. This applies to software modules as well. They should look nice, clear to understand, and possibly have a touch of craze together with a "trademark" color scheme or a logo that makes all the modules from the same developer instantly recognizable. Ugly, uninspiring panels are often disregarded, especially in the Rack third-party plugins list, where there's already so much stuff. A good mix of creativity and design experience makes the GUI catchy and inspiring, improving the user experience (UX).

In this chapter, we will deal with the practical details of creating a user interface. With Rack, the user interface is created in two stages:

1. Graphical design of the panel background (SVG file) using vector drawing software.
2. Code development to add components and bind them to parameters and ports of the module.

The SVG file is used as a simple background for the panel. All the interactive parts, such as knobs, connectors, and so on, are added in C++. You will be guided in the design of the background file in Section 5.1.

As reported in Section 4.5.2, the whole module is contained into a `ModuleWidget` subclass widget that collects all the "physical" elements (knobs, inputs, outputs, and other graphical widgets) and links them to the signals in the `Module` subclass. The widget class fills the front panel with the dynamic components that are required for interaction or graphical components that may change over time (e.g. a graphic display, an LCD display, or an LED). There is a fair choice of components that you can draw from the Rack APIs (see Section 5.2), but in order to design your customized user interface you will need to create new widgets. This last topic is postponed to Chapter 9.

5.1 Generating SVG Files with Inkscape

All modules require one background image, stored as a scalable vector graphics (SVG) file. The file is stored in the `ModuleWidget` subclass. SVG is an open-standard vector graphic format required to obtain the best results in terms of graphical rendering with different screen resolutions. A vector graphic file describes the graphics in terms of shapes and vector, while a raster graphic file describes the graphic in terms of pixels. Any graphic file in VCV Rack will require different rescaling for each different monitor or screen resolution. Alternatively, doing this with raster graphic formats such as PNG or JPG requires interpolation, introducing an inherent quality reduction and reducing the readability of the front panel texts.

SVG files can be drawn with several vector graphics applications. However, we will show how to do that using an open-source software, Inkscape. Inkscape is available for all platforms and generates SVG files that are ready to use with Rack. It can be downloaded from https://inkscape.org/ or from the software store available with some operating systems.

We will refer to the graphical design of the ABC plugins. These are extremely simple and clean in their aspect to make the learning curve quick. The ABC plugins adopt the following graphical specifications[1]:

- Light gray background fill with full opacity (RGBA color: 0xF0F0F0FF).
- A small 5 px-high blue band on bottom (RGBA color: #AACCFFFF).
- Title on top using a "sans serif" font (DejaVu Sans Mono, in our case, which is open-source) with font size 20 (RGBA color: #AACCFFFF).
- Minor texts (ports and knobs) will be all uppercase with DejaVu Sans Mono font but font size 14.
- Major texts (sections of a module, important knobs or ports) will be white in a blue box (RGBA color: #AACCFFFF).

5.1.1 Creating the Panel and the Background

First, we need to create a new file in Inkscape (File → New) and set some of its properties. Go to File → Document Properties (or Shift+Ctrl+D) and:

- Make "mm" the Default Units.
- Make "mm" the Units (under Custom Size line).
- Set the Height to 128.5 mm and the Width to a multiple of 5.08 mm (1 HP), according to the width of the module you want to create.

It is suggested to use a grid for better alignment, snapping, and positioning of objects. In the Document Properties, go to the Grid tab, create a New Rectangular Grid, and verify that the spacing for both X and Y is 1 px and the origin is at 0, 0. Checking the "Show dots instead of lines" box is suggested for improved visibility.

Now we have an empty page whose limits are those of our module front panel. We can now create the background as a rectangle (shortcut: F4) of size equal to the page size. *Hint*: In order to avoid displacement errors, we suggest using snapping, which automatically snaps the rectangle corners to the page corners. To do this, enable the following options: "Enable Snapping," "Snap bounding boxes," "Snap bounding box corners," and "Snap to the page border" (on the "Snap Control" bar).

We can open the "Fill and Stroke" dialog (shortcut: Ctrl+Shift+F) to set the color of the box. You can just input the RGBA color code that we decided at the beginning of this chapter. We also remove any stroke.

Finally, we decided to add a small colored band at the bottom of the panel as our "trademark," and to make the panel a bit less dull. We create another rectangle (shortcut: F4) and we snap it to the bottom of the page. This time in the "Fill and Stroke," we set a nice blue color as in the RGBA color code we decided at the beginning of this chapter.

The empty panel looks like Figure 5.1 now.

Figure 5.1: **An empty panel of width 12 HP designed with Inkscape.**

5.1.2 Creating the Panel and the Background

Currently, Rack does not support on-the-fly rendering of the texts incorporated in an SVG file. This is mainly due to limitations of the external library used for vector graphics rendering. If you place texts in an SVG file, the resulting module will not show them up once loaded into Rack. The proper way to add text to the module SVG file is to transform it into a regular path object. However, once transformed, you won't be able to change the text, and you will have to create a new one, replace the previous one, and transform into a path (see Figure 5.2).

To introduce a text in Rack, use the text object icon (shortcut: F8) and insert text. You can edit the font and other text properties using the text toolbar. After your text is ready and its graphical properties are OK, you need to transform to a path object. Save a copy of the file with all the text objects or duplicate the text object and place it somewhere outside the module box. Select the text

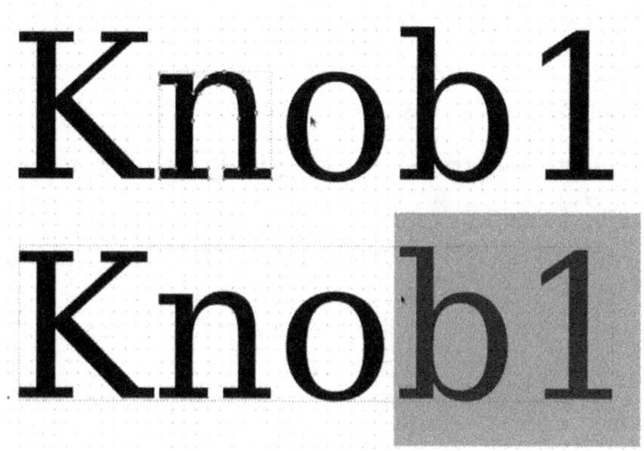

Knob1
Knob1

Figure 5.2: Comparison of a path object (top) and a text object (bottom). A path object is a vector object defined by nodes, which can be edited. A text object, instead, can be edited as ... well ... text, with all the font rendering options and so on.

object and transform it using the menu Path → Object to Path. Now the text cannot be edited anymore.

This may get annoying, as you always need to keep for yourself a version of the module with the texts and another with the paths to allow for later editing whenever you change your mind about the name of a knob or its appearance.

For the ABC modules, we will often print text directly from the code. This practice should be avoided for released modules as it adds some overhead, but facilitates the development cycle.

5.2 Base Widget Objects for Knobs, Ports and Switches

We will survey here the objects available that can populate a module panel. All the objects are of the `struct` type and they descend from a few base classes:

- `SvgKnob`. Defines a rotating knob.
- `SvgPort`. Defines an input or output port.
- `LightWidget`. An LED widget that can have one or multiple colors, depending on the value we assign to it.
- `SvgSwitch`. Defines a switch; it must be inherited together with one of the two classes below.

These classes are inherited to get diverse results. For instance, the class RoundBlackSnapKnob inherits from the following hierarchy:

- `SvgKnob`
  - `RoundKnob`. Defines minimum and maximum rotation angles.
    - `RoundBlackKnob`. Defines the SVG image, thus imposing the size.
      - `RoundBlackSnapKnob`. Defines the snap property true, making the knob snap on integer values.

From this hierarchy, the last two are usable in your designs, while the first two are made only to be inherited as they will not show up on your module, because they don't define an SVG image file for rendering.

The file include/componentlibrary.hpp defines all the available objects.

To include one of the components widgets in your ModuleWidget, you need to add it according to the methods described below.

The syntax for an input or output port is:

```
addInput(Port::create<port-type>(position, Port::INPUT, module,
number));
addOutput(Port::create<port-type>(position, Port::OUTPUT, module,
number));
```

where:

- Port-type is the name of the widget port class we want to instantiate (e.g. one of the classes for inputs and outputs available in include/componentlibrary.hpp, such as `PJ301MPort`). This basically defines how the port will look like.
- Position is a `Vec(x,y)` instance telling the position of the top-left corner of the widget, in pixels.
- Module is the module class pointer; you usually don't have to change this.
- *Number* is the index of the related input our output according to the `InputIds` or `OutputIds` enum, indicating which signal is generated from this input (or which signal is sent to this output). The signals are accessible with `inputs[number].value` or `outputs[number].value`, as we shall see later. However it is generally safer to use the name from the enum (e.g. `myModule::MAIN_INPUT`) so that you don't risk messing up with the numbers.

The syntax for a parameter (knob, switch, etc.) follows:

addParam(ParamWidget::*create<class-type>*(position, module, *number*, min, max, default));

where:

- Class-type is the name of the class we want to instantiate (e.g. one of the classes for knobs and buttons seen in include/componentlibrary.hpp, such as RoundBlackKnob or NKK for a knob or a type of toggle switch).
- Position is a `Vec(x,y)` instance telling the position of the top-left corner of the widget, in pixels.
- Module is the module class pointer; you usually don't have to change this.
- *Number* is the index of the related parameter according to the `ParamIds` enum, indicating which parameter value is generated when the user interacts with this object. The value generated by this object is accessible with `params[number].value`, as we shall see later. However, it is generally safer to use the name from the enum (e.g. `myModule::CUTOFF_ PARAM`) so that you don't risk messing up with the numbers.
- Min, max, default: these three values indicate the range of the knob/button (min to max) and the default value assumed by the button. For knobs, it may be any floating-point value. For buttons, it should be something like 0.f and 1.f as min and max and 0.f as default.

The syntax for a light is as follows:

```
addChild(ModuleLightWidget::create<size-type<color-type>>(position,
module, number));
```

where:

- Size-type is the name of a class that defines the light size (and other characteristics too). It is one of the classes in include/componentlibrary.hpp, such as `LargeLight`, `MediumLight`, `SmallLight`, and `TinyLight`. All these classes require a template argument (see below).
- Color-type is the name of a class that defines the color of the light, such as `GreenLight`, `RedLight`, `YellowLight`, `GreenRedLight`, and so on. This is the template argument for the size-type. Each of these objects defines at least one color for the light.
- Position is a `Vec(x, y)` instance telling the position of the top-left corner of the widget, in pixels.
- Module is the module class pointer; you usually don't have to change this.
- *Number* is the index of the related light according to the `LightsIds` enum, indicating which light is this. The value that determines the status and color of the light is accessible with `lights[number].value`, as we shall see later. However, it is generally safer to use the name from the enum (e.g. `myModule::ACTIVE_LIGHT`) so that you don't risk messing up with the numbers.

5.2.1 *Automatically Placing Components*

Adding components requires you to know beforehand *where* to place them and providing the coordinates in pixels. This, however, requires an initial guess, and some iterations of trial and error, in order to refine the coordinates. This can get particularly hard if the component has to fit a specific position in the SVG panel background.

The easiest solution is to employ a helper script provided by Rack that converts the SVG file into a template C++ file. The script prepares the whole skeleton for the Module and the ModuleWidget, adds components in place, and also modifies the JSON manifest. The script, helper.py, is located in the Rack root. It places components according to this convention: it looks for an optional layer in your SVG file where you can draw placeholders and converts them to C++ GUI objects according to their color code, position, and shape. This layer must be named components and should be set to invisible, to avoid the placeholders showing up when the module is loaded. The placeholders are converted into Rack components according to their color:

- red (RGBA: #FF0000) for Parameters (knobs, switches, sliders);
- green (RGBA: #00FF00) for Input ports;
- blue (RGBA: #0000FF) for Output ports;
- magenta (RGBA: #FF00FF) for Lights; and
- yellow (RGBA: #FFFF00) for custom widgets of yours.

The placing of components may be relative to the center or the top-left corner – draw a circle (F5) for the former and a rectangle (F4) for the latter. The placement in C++ corresponds to assigning horizontal and vertical coordinates using a Vec object, which is initialized as `Vec(<horizontal>, <vertical>)`. The name of the object may be given directly in the SVG file to avoid unnecessary editing later in the code. Open the "Object Properties" panel (Shift+Ctrl+O), select a placeholder shape, and type in the "ID" field. Please remember to press "Set" to apply the new name.

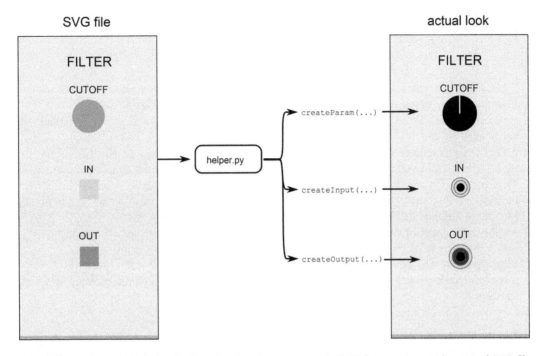

Figure 5.3: Using coloured placeholders (circle and squares on the left) for components in a panel SVG file. The helper.py script generates code for the ModuleWidget struct, analyzing the shape type, position, and color of the placeholders in the "components" layer of the SVG file. The final rendering is done by Rack when the compiled module is loaded.

An explanatory example is provided in Figure 5.3, where three colored shapes in the SVG are translated to the C++ `createParam`, `createInput`, and `createOutput` commands that draw the respective components during the in-application rendering. Let us call them, from top to bottom, CUTOFF, FILTER_IN, and FILTER_OUT by setting their ID in the SVG. We save the file in Rack/plugins/myPlugin/res/myModule.svg. By issuing the following commands:

```
cd Rack/plugins/myPlugin/
```

```
../../helper.py createmodule myModule <svg file> <destination cpp file>
```

the helper script is invoked. It will ask us a few questions to compile the manifest and give a name to the module, then it will compile the manifest with the given data and generate source code. The output of the script will state if an SVG panel was found and how many placeholders were found:

```
Added myModule to plugin.json
```

```
Panel found at res/myModule.svg. Generating source file.
```

```
Found 1 params, 1 inputs, 1 outputs, 0 lights, and 0 custom widgets.
```

```
Components extracted from res/myModule.svg
```

```
Source file generated at src/myModule.cpp
```

Taking a look at myModule.cpp, we see the following lines:

```
struct MyModule: Module {
  enum ParamIds {
      CUTOFF_PARAM,
      NUM_PARAMS
  };
  enum InputIds {
      FILTER_IN_INPUT,
      NUM_INPUTS
  };
  enum OutputIds {
      FILTER_OUT_OUTPUT,
      NUM_OUTPUTS
  };
  enum LightIds {
      NUM_LIGHTS
  };

  MyModule() {
      config(NUM_PARAMS, NUM_INPUTS, NUM_OUTPUTS, NUM_LIGHTS);
      configParam(CUTOFF_PARAM, 0.f, 1.f, 0.f, "");
  }

  void process(const ProcessArgs &args) override {
  }
};

struct MyModuleWidget: ModuleWidget {
  MyModuleWidget(MyModule *module) {
      setModule(module);
      setPanel(APP->window->loadSvg(asset::plugin(pluginInstance,
"res/myModule.svg")));

      addChild(createWidget<ScrewSilver>(Vec
(RACK_GRID_WIDTH, 0)));
      addChild(createWidget<ScrewSilver>(Vec(box.size.x - 2 *
RACK_GRID_WIDTH, 0)));
      addChild(createWidget<ScrewSilver>(Vec(RACK_GRID_WIDTH,
RACK_GRID_HEIGHT - RACK_GRID_WIDTH)));
      addChild(createWidget<ScrewSilver>(Vec(box.size.x - 2 *
RACK_GRID_WIDTH, RACK_GRID_HEIGHT - RACK_GRID_WIDTH)));

      addParam(createParam<RoundBlackKnob>(mm2px(Vec(23.666,
22.16)), module, MyModule::CUTOFF_PARAM));

      addInput(createInput<PJ301MPort>(mm2px(Vec(26.615, 56.027)),
module, MyModule::FILTER_IN_INPUT));
```

```
        addOutput(createOutput<PJ301MPort>(mm2px(Vec(26.615,
84.506)), module, MyModule::FILTER_OUT_OUTPUT));
        }
};

Model *modelMyModule = createModel<MyModule, MyModuleWidget>
("myModule");
```

The skeleton of the module is ready. The components have been placed with the `createParam`, `createInput`, and `createOutput` methods, and the names of the components are those set in the SVG file. Positions are given using `Vec` objects and converting the values from mm (as used in the SVG) to px (as used in the widget). Most importantly, these names are also used in the `myModule` struct, in the related enums, in the `config` and `configParam` function calls. At this point, one can start coding and tweaking the GUI by modifying the components.

Note

1 Please note that colors are expressed as RGBA, which stands for red green blue alpha (i.e. transparency). Please bear in mind that an alpha value of 0 corresponds to a totally transparent color, thus invisible, while an alpha value of 255 corresponds to no transparency. The 8-bit values (0–255) are provided in hexadecimal notation.

CHAPTER 6

Let's Start Programming: The Easy Ones

The boring part is over; from here on. we are going to build and test modules!

If you followed the previous chapters, you are already able to compile a module and edit its graphical user interface. In this chapter, we start with coding. The easiest way to take acquaintance with coding with VCV Rack is to build "utility" modules that are functional but do not require extensive DSP knowledge or complex APIs. We are going to set up a comparator module, a multiplexer/demultiplexer (in short, mux and demux), and a binary frequency divider, shown in Figure 6.1.

6.1 Creating a New Plugin from Scratch, Using the Helper Script

Before we start with the modules, we need to create the basics for a new plugin. This will allow us to add modules one at a time.

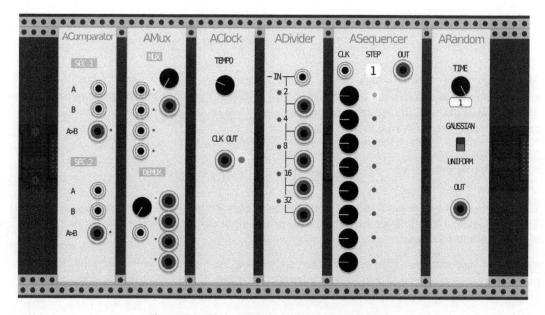

Figure 6.1: The modules presented in this chapter.

The basic steps follow:

- Create a new folder in Rack/plugins/.
- Create the folders for the source code and the resources, src/ and res/.
- Generate a plugin.json manifest.
- Add a license file and a readme (not strictly necessary, but useful for sharing with others).
- Add a Makefile.
- Add the C++ and header files for the plugin.

To get this work sorted out quickly and free of errors, the Rack helper script under Rack/helper.py again comes in handy.

Go to Rack/plugins/ and invoke it:

```
cd Rack/plugins/

../helper.py createplugin myPlugin
```

An interactive prompt will ask for all the details that will go into the JSON manifest. A folder including all the necessary files will be created, leaving you ready for the task.

During the rest of the book, we shall work on the development of the ABC plugin collection, thus the plugin name will be ABC, and all the modules will start with the letter "A" (e.g. AComparator, etc.).

6.2 Comparator Module

 TIP: In this section, we will discuss the whole code and process involved in creating your first complete module.

As a first example, we start with a simple module, a comparator. Given two input voltages, a comparator module simply provides a high output voltage if input 1 is higher than input 2, or a low voltage otherwise. In our case, the high and low voltages are going to be 10 V and 0 V. The module we are going to design has:

- two inputs;
- one output; and
- one light indicator.

These are handled by the process(...) function, which reads the input and evaluates the output value. The light indicator also follows the output. We shall, first, describe the way inputs, outputs, and lights are handled in code.

Let us first concentrate on monophonic inputs and outputs. As discussed in Section 4.1.3, Rack provides a fancy extension to real-world cables (i.e. polyphonic cables), but for the sake of simplicity we shall leave these for later.

Input and output ports are, by default, monophonic. Three methods allow you to access basic functionalities valid for all type of ports:

- `isConnected();`
- `getVoltage();` and
- `setVoltage(float voltage).`

The first method allows you to check whether a cable is connected to the port. The getVoltage() method returns the voltage applied to the port. This is normally used for reading the value of an input port. Finally, the last method sets the voltage of the port (i.e. assigns a value to an output port). These three methods are enough for most of the functionalities we shall cover in this book.

The lights have their own set/get methods:

- `getBrightness();` and
- `setBrightness(float brightness).`

These clearly have similar functionalities; however, we are going to use the latter most of the time to assign a brightness value to the lights.

While the port voltage follows the Eurorack standards (see Section 1.2.3), the lights get a brightness value in the range [0, 1]. The value is squared, so negative values will be squared too, and larger values are clipped.

Now that you have all the necessary bits of knowledge, we can move to the development of the comparator module. The development steps (not necessarily in chronological order) follow:

- Prepare all the graphics in res/.
- Add the module to plugin.json.
- Add the C++ file AComparator.cpp to src/.
- Develop a struct (Acomparator) that subclasses Module.
- Develop a struct (AComparatorWidget) that subclasses ModuleWidget and its constructor.
- Create a model pointer modelAComparator that takes the above two structs as template arguments.
- Add the model pointer to the plugin into ABC.cpp.
- Declare the model pointer as extern in ABC.hpp.

Naturally, the helper script may speed up the process.

We will start defining how the module behaves. This is done by creating a new struct that inherits the Module class.

Let's take a look at the class definition, which shall be placed in AComparator.cpp:

```
struct AComparator : Module {
enum ParamIds {
    NUM_PARAMS,
};
enum InputIds {
    INPUTA1,
    INPUTB1,
    INPUTA2,
    INPUTB2,
```

```
            NUM_INPUTS,
    };
    enum OutputIds {
            OUTPUT1,
            OUTPUT2,
            NUM_OUTPUTS,
    };

    enum LightsIds {
            LIGHT_1,
            LIGHT_2,
            NUM_LIGHTS,
    };

    AComparator() {
            config(NUM_PARAMS, NUM_INPUTS, NUM_OUTPUTS, NUM_LIGHTS);
    }

    void process(const ProcessArgs &args) override;
};
```

The module defines some enums that conveniently provide numbering for all our inputs, outputs, and lights. You may want to give meaningful names for the enums. Always remember that, generally, enums start with 0.

We don't have members for this module, so we go straight into declaring the methods: we have a constructor, `AComparator()`, and a `process(...)`, which will implement the DSP stuff. That's all we need for this basic example.

The constructor is the place where we usually allocate stuff, initialize variables, and so on. We leave it blank for this module and go straight into the process function.

The `process(...)` function is the heart of your module. It is called periodically at sampling rate, and it handles all the inputs, outputs, params, and lights. It implements all the fancy stuff that you need to do for each input sample.

Each time the process function is executed, we compare the two inputs and send a high-voltage or a low-voltage value to the output. This is done by a simple comparison operator.

In C++ code:

```
void AComparator::process(const ProcessArgs &args) {

    if (inputs[INPUTA1].isConnected() && inputs[INPUTB1].
isConnected())
    {
        float out = inputs[INPUTA1].getVoltage() >= inputs[INPUTB1].
getVoltage();
            outputs[OUTPUT1].setVoltage(10.f * out);
            lights[OUTPUT1].setBrightness(out);
    }
}
```

We want to spare computational resources and avoid inconsistent states, thus sometimes it is better to avoid processing the inputs if they are not "active" (i.e. connected to any wire). If they are both connected, we evaluate their values and use the operator "?" to compare the two. The result is a float value that takes the values of 0.f or 1.f. The output goes to the light (so we have visual feedback even without an output wire) and to the output voltage.

Now that the module guts are ready, let's get started with the widget. We declare it in AComparator. cpp. This widget is pretty easy and it only requires a constructor to be explicitly defined.

```
struct AComparatorWidget : ModuleWidget {

AComparatorWidget(AComparator* module) {

    setModule(module);
    setPanel(APP->window->loadSvg(asset::plugin(pluginInstance,
        "res/AComparator.svg")));

    addInput(createInput<PJ301MPort>(Vec(50, 78), module,
        AComparator::INPUTA1));
    addInput(createInput<PJ301MPort>(Vec(50, 108), module,
        AComparator::INPUTB1));

    addOutput(createOutput<PJ3410Port>(Vec(46, 138), module,
        AComparator::OUTPUT1));

    addChild(createLight<TinyLight<GreenLight>>(Vec(80, 150),
        module, AComparator::LIGHT_1));
    }

};
```

What is the constructor doing? It first tells `AComparatorWidget` to take care of AComparator by passing the pointer to the `setModule(...)` function. It then sets the SVG graphic file as the panel background. The file is loaded as an asset of the plugin and automatically tells Rack about the panel width.

At the end of the widget, we add other children as the inputs and outputs. Let's have a detailed look into this. The `addInput` method requires a `PortWidget*`, created using the createInput method. This latter method requires a port template, which basically defines the graphical aspects (as far as we are concerned now) of this new port. The creation argument of the port defines the position, in terms of a (x,y) coordinate vector Vec, the parent module, and the port number of that module. We use, for convenience and readability, the enum name instead of a number. Similarly, we do for the output ports and the parameters (although we do not have them in this example). Other widgets, such as lights and screws,[1] use an `addChild` method.

The widget configuration, as you can see, is pretty basic, and most modules will follow the same guidelines:

```
Model * modelAComparator = createModel<AComparator,
AComparatorWidget>("AComparator");
```

Finally, the Model is created, setting the Module subclass and the ModuleWidget subclass as template arguments and the module slug as input argument.

The last step is to create this model and add it to the list of models contained in our plugin. We do this in ABC.cpp inside the **init**(rack::Plugin *p) function as:

```
p->addModel(modelAComparator);
```

and we declare this model pointer as extern in ABC.hpp, so that the compiler knows that it exists:

```
extern Model * modelAComparator;
```

OK, pretty easy! We can compile the code against the provided version of the VCV Rack source code, *et voila*, the module is ready to be opened in VCV Rack!

Now it is time for you to test your skills with some exercises. You will find this simple code in the book repository. Go to the exercise section of this chapter to see how to expand the module and improve your skills!

Exercise 1

The comparator module has only two inputs and an output port. Wasting space is a pity, so why don't we replicate another comparator so that each module has two of them? You can now try to add two extra inputs and one extra output and light. You can double most parts of the code, but I suggest you iterate with a for loop through all the inputs and outputs. Try on your own, and at the end – if you can't figure it out on your own – take a look at how this is done in the AComparator found with the online resources.

Exercise 2

Sometimes you want the output to latch for a few milliseconds (i.e. to avoid changing too quickly). Think of the case where input "A" is a signal with some background noise and input "B" is zero. All the times that "A" goes over zero, the output will get high. This will happen randomly, following the randomness of the noisy input "A" signal. How do you avoid this? A hysteresis system is what you need. Try to design yourself a conditional mechanism that:

- drives the output high only when input "A" surpasses a threshold that is slightly higher than input "B"; and
- drives the output low only when input "A" drops below a threshold that is slightly lower than input "B."

6.3 Muxing and Demuxing

 TIP: In this section, you will learn how to add parameters such as knobs, and to pick the right one from the component library.

The terms "mux" and "demux" in engineering stand for multiplexing and demultiplexing, which are the processes that make multiple signals share one single medium, in our case a wire (or its virtual counterpart). In analog or virtual circuits, we need multiplexers to allow two or more signals to be transferred over a wire, one at a time. Similarly, we demultiplex a wire when we redirect it to multiple outputs. This is achieved in analog circuits by means, for example, of voltage-controlled switches. In the digital domain, we do it a bit differently.

Considering an integer variable, selector, holding the value of the input port to be muxed to the output, a C++ snippet for a multiplexer could be as follows:

```
output = inputs[selector]; // MUX
```

Similarly, a C++ snippet for a demultiplex, where one input can be sent to one of many outputs, looks like:

```
outputs[selector] = input; // DEMUX
```

How do we translate this into Rack port APIs? Try this on your own first – write it on paper, as an exercise.

This second module we are designing requires a selector knob. This is a parameter, in the Rack terminology. Let us discuss how parameters work and how to add them to a module. Parameters may be knobs, sliders, switches, or any other graphical widget of your invention that can store a value and be manipulated by the user. The two important methods to know are:

- `getValue();` and
- `setValue(float value)`.

As with lights and ports, the set/get methods return or set the value of the parameter. While the latter is of use in some special cases (e.g. when you reset or randomize the parameter), you will use the `getValue()` most of the time to read any change in the value from the user input.

The parameters are added to the `ModuleWidget` subclass to indicate their position and bind them to one of the parameters in the enum ParamIds. Let us look at this example:

```
addParam(createParam<ParameterType>(Vec(X, Y), module, MyModule::
OUTPUT_GAIN));
```

The details you have to input in this line are the `ParameterType` (i.e. the name of a class defining the graphical aspect and other properties of the object, the positioning coordinates X and Y, and the parameter from the enum ParamIds of the Module struct to which you bind this parameter).

The last bit of information to give to the system is the value mapping and a couple of strings. All these are set through the configParam method, which goes into the Module constructor. One example follows:

```
configParam(OUTPUT_GAIN, 0.f, 2.f, 1.f, "Volume");
```

In this case, we are telling the Module that the knob or slider related to the parameter OUTPUT_GAIN should go from 0 to 2, and by default (initialization) will be 1. The string to associate to it when right-clicking is "Volume." Further arguments allow you to make this more powerful, but we'll see this later in this chapter.

Now let us move through the implementation of the whole module, AMuxDemux. The module will host both mux and demux sections, and thus two selector knobs will be necessary. We shall have four inputs for the mux and four outputs for the demux. We shall also place light indicators near each of the selectable inputs for the mux, and near each of the selectable outputs for the demux.

In our case, if we define the enums as:

```
enum ParamIds {
      M_SELECTOR_PARAM,
      D_SELECTOR_PARAM,
      NUM_PARAMS,
};
enum InputIds {
      M_INPUT_1,
      M_INPUT_2,
      M_INPUT_3,
      M_INPUT_4,
      D_MAIN_IN,
      NUM_INPUTS,
      N_MUX_IN = M_INPUT_4,
};
enum OutputIds {
      D_OUTPUT_1,
      D_OUTPUT_2,
      D_OUTPUT_3,
      D_OUTPUT_4,
      M_MAIN_OUT,
      NUM_OUTPUTS,
      N_DEMUX_OUT = D_OUTPUT_4,
};

enum LightsIds {
      M_LIGHT_1,
      M_LIGHT_2,
      M_LIGHT_3,
      M_LIGHT_4,
      D_LIGHT_1,
      D_LIGHT_2,
      D_LIGHT_3,
      D_LIGHT_4,
      NUM_LIGHTS,
};
```

we declare two integer variables to hold the two selector values. To make their values consistent across each step of the process(…) method, we declare them as members of the AMuxDemux struct:

```
unsigned int selMux, selDemux;
```

They are zeroed in the Module constructor, where the value mapping of the selectors is also defined:

```
AMuxDemux() {
    config(NUM_PARAMS, NUM_INPUTS, NUM_OUTPUTS, NUM_LIGHTS);
    configParam(M_SELECTOR_PARAM, 0.0, 3.0, 0.0, "Mux Selector");
    configParam(D_SELECTOR_PARAM, 0.0, 3.0, 0.0, "Demux Selector");
    selMux = selDemux = 0;
}
```

We can now implement the process method as follows:

```
void AMuxDemux::process(const ProcessArgs &args) {

/* MUX */
lights[selMux].setBrightness(0.f);
selMux = (unsigned int)clamp((int)params[M_SELECTOR_PARAM].
    getValue(), 0, N_MUX_IN);
lights[selMux].setBrightness(1.f);

if (outputs[M_MAIN_OUT].isConnected()) {
    if (inputs[selMux].isConnected()) {

outputs[M_MAIN_OUT].setVoltage(inputs[selMux].getVoltage());
    }
}

/* DEMUX */
lights[selDemux+N_MUX_IN+1].setBrightness(0.f);
selDemux = (unsigned int)clamp((int)params[D_SELECTOR_PARAM].
    getValue(), 0, N_DEMUX_OUT);
lights[selDemux+N_MUX_IN+1].setBrightness(1.f);

if (inputs[D_MAIN_IN].isConnected()) {
    if (outputs[selDemux].isConnected()) {

outputs[selDemux].setVoltage(inputs[D_MAIN_IN].getVoltage());
    }
}
}
```

As you can see, there are several checks and casts. We need to cast the input value to an integer because it is used as an index in the array of inputs. The selector values are always clamped using Rack function clamp(), which clamps a value between a minimum and a maximum. This is a little paranoid, but it is very important that the selector value does not exceed the range of the arrays they index, otherwise an unpleasant segmentation fault is issued and Rack crashes. You can avoid this if you make the parameters D_SELECTOR_PARAM and M_SELECTOR_PARAM be constrained properly in your widget.

You will notice that we shut off a light each time before evaluating the new value for selDemux or selMux, even if the selector has not changed. This makes the code more elegant and reduces the number of *if* statements.

Finally, the lines concerning the outputs are reached only if the relevant inputs and outputs are connected. The assignments to the output ports are done using the setVoltage() and getVoltage() methods. Compare this with the lines you wrote previously for the exercise. Did you get the thing right?

To conclude the module, let us look at the graphical aspect. We defined through the configParam method that the selectors span the range 0–3. The last step is to tell the ModuleWidget subclass where to place the selector, and what it looks like. The latter is defined by the template <ParameterType> of the createParam method, seen above. We have a lot of options – just open the include/componentlibrary.hpp and scroll. This header file defines a lot of components: lights, ports, and parameters. Let us look at the knobs: RoundKnob, RoundLargeBlackKnob, Davies1900hWhiteKnob, and so on – there are a lot of options! As discussed in Section 5.2, they are organized hierarchically, by inheritance, and their properties are easy to interpret from the source code directly.

Besides the graphical aspect, there is one important property to make the mux/demux module work flawlessly: the snap property. Knobs can snap to integer values, as a hardware selector/encoder would do, skipping all the real-valued position between two integer values. This way, the user can select which of the input/output ports to mux/demux easily, with a visual feedback. We take the RoundBlackSnapKnob from the component library and place it where it fits as follows:

```
addParam(createParam<RoundBlackSnapKnob>(Vec(50, 60), module,
AMuxDemux::M_SELECTOR_PARAM));
```

Before moving on to the last remarks, take a look at the provided code, AMuxDemux.cpp. The module implements both a multiplexer (upper part) and a demultiplexer (bottom part), so you can test it easily.

Note: The function clamp is an overloaded function. This means that there are two definition of the function, one for integer and one for float variables. You can use the same function name with either float or integer values, and the compiler will take care of figuring out which one of the two implementations to use, depending on the input variable type.

Exercise 1

What about employing an input to select the output (input) of a multiplexer (demultiplexer)? This is handy as we may want to automate our patches with control voltages. To implement this, it is sufficient to replace the param variable with an input variable.

Exercise 2

You may decide to have both a knob and an input as a selector. How could you make the two work together? There is no standard for this, and we may follow different rules:

- Sum the input CV and the knob values after casting to int: this is quite easy to understand for the user. It may not be of use if the knob is turned all the way up: the input signal will not affect the behavior in any way unless it is negative.
- Sum the input and the knob values and apply a modulo operator: this way, there will always be a change, although it may be more difficult to have control on it.
- Exclude the knob when the CV input is active.

6.4 Knobs: Displaying, Mapping, and Computing Their Values

Before going on with the development of modules, we should focus on knobs, or more generally on parameters. The configParam was introduced in Section 4.3; however, now it is time to discuss it in more detail. The arguments it takes are:

1. `paramId`. The parameter we are configuring.
2. `minValue`. The minimum value.
3. `maxValue`. The maximum value.
4. `defaultValue`. The default value for initialization or reset.
5. `label`. The name that is visualized.
6. `unit`. The measurement unit (if any) to be displayed.
7. `displayBase`. Used to compute the displayed value.
8. `displayMultiplier`. Multiplies the displayed value.
9. `displayOffset`. Adds an offset to the displayed value.

The last three arguments can be important to improve the user experience: they are meant to compute a value that is displayed to the user in order for the knob to make sense. This value is displayed in the parameter tooltip or when right-clicking on the parameter. Depending on displayBase, the way the displayed value is computed changes:

$$
\begin{cases}
v \cdot m + ob = 0 & (6.1a) \\
\frac{\log(v)}{\log(-b)} \cdot m + ob < 0 & (6.1b) \\
b^v \cdot m + ob > 0 & (6.1c)
\end{cases}
$$

where b, m, and o are `displayBase`, `displayMultiplier`, and `displayOffset`, respectively, and v is the value given by the knob, in the range `minValue` to `maxValue`.

When the last three arguments are discarded, the parameters scroll linearly from the `minValue` to `maxValue`. When the base b is zero but any of m and/or o are provided, the value is computed according to Equation 6.1a (i.e. a multiplication factor and an offset can be applied). If the base is negative, the logarithmic case applies, as in Equation 6.1b. The natural logarithm of $-b$ (to make it positive) is used to scale the natural logarithm of the knob value. A multiplier and an offset are still applied after computing the logarithms. Please note that the denominator should not be zero, thus $b \neq -1$. Finally, if the base is positive, the base is raised to the power of v.

Examples of these mappings are shown in Figure 6.2.

Please note that the displayed value is computed by the GUI in Rack only for display. The value provided by the `Param::getValue` method will always be v (i.e. the linear value in the range from minValue to maxValue). This leaves you free to use it as it is, to convert it in the same way it is computed according to the configParam, or even compute in a different way. Considering the computational cost of transcendental functions such as `std::pow` or `std::log`, you should

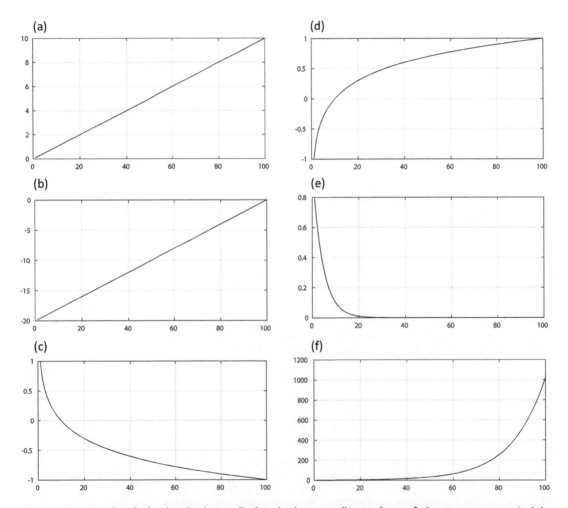

Figure 6.2: Mapping the knob value into a displayed value according to the configParam arguments. In (a), the values v, ranging from 0.1 to 10, are shown. In (b), the mapping of Equation 6.1a is shown with $m = 2, o = -20$. In (c) and (d), the mappings are shown according to Equation 6.1b for $b = 0.1$ and $b = 10$, respectively. In (e) and (f), the exponentials are shown with $b = 0.1$ and $b = 2$.

avoid evaluating the value according to Equation. 6.1b and Equation 6.1c at each time step. Good approximations of logarithms can be obtained by roots. Roots and integer powers of floats are quicker to compute than logarithms or float powers.

In the following chapters, different strategies will be adopted to map the parameters according to perception and usability.

6.5 Clock Generator

TIP: In this section, you will learn how to generate short clock pulses using the PulseGenerator Rack object.

A clock generator is at the heart of any modular patch that conveys a musical structure. The notion of a clock generator differs slightly among the domains of music theory, computer science, communication, and electronics engineering. It is best, in this book, to abstract the concept of clock and consider it as a constant-rate events generator. In the Eurorack context, an event is a trigger pulse of short duration, with a voltage rising from low to high voltage. In VCV Rack, the high voltage is 10 V and the pulse duration is 1 ms.

In this section, we will develop a simple clock that generates trigger pulses at a user-defined rate expressed in beats per minute (bpm). The generated pulses can be used to feed a frequency divider or an eight-step sequencer as those developed in this chapter.

The operating principle of a digital clock is simple: it takes a reference clock at frequency f_{ref} and scales it down by issuing an event every L pulses of the reference clock. The obtained frequency is thus $f_{clk} = \frac{f_{ref}}{L}$. The reference clock must be very stable and reliable, otherwise the obtained clock will suffer jitter (i.e. bad timing of its pulses). In the hardware domain, a reference clock is a high-frequency PLL or quartz crystal oscillator. In the software domain, a reference clock is a reliable system timer provided by the operating system or by a peripheral device. The clock generator will count the number of pulses occurring in the reference clock and will issue a pulse itself when L pulses are reached.

As an example, consider a reference clock running at 10 kHz. If I need my clock generator to run at 5 kHz, I will make it issue a pulse every second pulse of the reference clock. Each time I pulse my clock, I also need to reset the counter.

In Figure 6.3, the frequency to be generated was an integer divisor of the reference clock frequency. However, when we have a clock ratio that is not integer, things change.

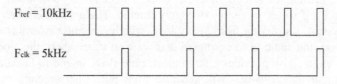

Figure 6.3: Generation of a 5 kHz clock from a 10 kHz clock by firing a pulse each second pulse of the reference clock.

Let us consider the following case. We need to generate a 4 kHz clock from a 10 kHz clock. The exact number of pulses to wait from the reference clock before firing a pulse is $L_{4k} = \frac{10\,kHz}{4\,kHz} = 2.5$, which is not integer. With the aforementioned method, I can only count up to 2 or 3 before issuing a pulse. If I fire a pulse when the counter reaches 3, I generate a clock that is 3.3 kHz, yielding an error of 0.6 kHz, approximately 666 Hz. Not very precise. Firing a pulse when the counter reaches 2 means generating a clock at 5 kHz, which is even farther from 4 kHz.

If I could increase the frequency of the reference clock, the error would be lower, even zero. If I have a clock at, for example, 100 kHz, I could count up to 25 and obtain a clock of 4 kHz. A larger clock ratio f_{ref}/f_{clk} is thus of benefit to reduce the error or even cancel it, where f_{ref} is a multiple of f_{clk}.

However, if I cannot change the ratio between the two clocks, a better strategy needs to be devised. In software, we can use fractional numbers for the counter. One simple improvement can thus be done by using a floating-point counter instead of an integer one. Let us again tackle the case where we want to generate a 4 kHz clock from a 10 kHz reference clock. As before, I can still fire a pulse only in the presence of a pulse from the reference clock, which is equivalent to setting an integer threshold for the counter, but this time I allow my counter to go below zero and have a fractional part.

The counter threshold $\hat{L}$ is set to 2 and the floating-point counter starts from zero. At the beginning, after counting two pulses of the reference clock, a pulse will be fired, but without resetting the counter to 0. Instead, L_{4k} is subtracted from the counter, making it go to -0.5. After two oscillations of the reference clock, the counter will reach 1.5, not yet ready to toggle. It will be toggled after the third reference pulse because the counter goes beyond 2 (more precisely, it reaches 2.5). At this point, L_{4k} is subtracted and the counter goes to 0, and we get back to the starting point. If you continue with this process, you will see that it fires once after two reference pulses and once after three reference pulses. In other words, the first pulse comes too early, while the second comes too late. Alternating the two, they compensate each other. If you look on a large time frame (e.g. over 100 reference clock pulses), you see 40 clock pulses (i.e. the average frequency of the clock is 4 kHz and the error – on average – is null). If you think about the whole process, it is very similar to the leap year (or bissextile year), where one day is added every four years because we cannot have a year that lasts for 365 days and a fraction.[2]

Please note that this way of generating a clock may not be desirable for all use cases, because the generated clock frequency is correct only on average. In fact, it slightly deviates from period to period, generating a jitter (i.e. an error between the expected time of the pulse and the time the pulse happens). In the previous example, the jitter was heavy (see Figure 6.4) because we had a very low clock ratio. In our scopes, however, the method is more than fine, and we will not dive into more complex methods to generate a clock. We are assuming that the clock ratio will always be large because we will be using the audio engine as a clock source to generate bpm pulses, which are times longer than the sample rate. The audio engine is very stable and reliable, otherwise the audio itself would be messy or highly jittered. The only case when the assumption fails is when the system is suffering from heavy load, which can cause overruns or underruns. However, in that case, the audio is so corrupted that we won't care about the clock generator module. The audio engine also provides a large clock ratio if we intend to build a bpm clock generator (i.e. a module that at its best will generate a few pulses per second). Indeed, the ratio between the reference clock (44,100 at lowest) and the bpm clock is much larger than the example above. While in the example above the reference clock was less than an order of magnitude faster,

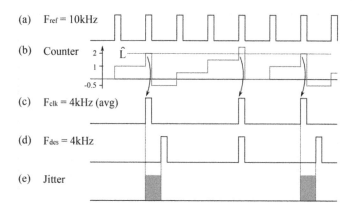

(a) F_{ref} = 10kHz

(b) Counter

(c) F_{clk} = 4kHz (avg)

(d) F_{des} = 4kHz

(e) Jitter

Figure 6.4: The process of generating a clock that has an average frequency of 4 kHz from a 10 kHz clock source. The figure shows the reference clock (a), the floating-point counter (b), the clock triggered by the counter (c), which will be 4 kHz only on average, the desired clock at precisely 4 kHz (d), and the time difference between the generated clock and the desired one (e). Although the generated clock has an average frequency of 4 kHz, it is subject to heavy jitter because of the low clock ratio.

in our worst case here we have a reference clock that is more than three orders of magnitude faster. If you take, for example, 360 bpm as the fastest clock you want to generate, this reduces to 6 beats per second, while the audio engine generates at lowest 44,100 events per second. The ratio is 7,350. In such a setting, the jitter cannot be high, and the average error, as shown above, reduces to zero.

Let us now see how to build the module in C++.

We first describe the PulseGenerator object. This is a struct provided by Rack to help you create standard trigger signals for use with any other module. The struct helps you generate a short positive voltage pulse of duration defined at the time of triggering it. You just need to trigger when needed, providing the duration of the pulse, and then let it go: the object will take care of toggling back to 0 V when the given pulse duration is expired. The struct has two methods:

- `process(float deltaTime);` and
- `trigger(float duration).`

The latter is used to start a trigger pulse. It takes a float argument, telling the duration of the pulse in seconds (e.g. 1e-3 for 1 ms). The process method must be invoked periodically, providing, as input argument, the time that has passed since the last call. If the pulse is still high, the output will be 1. If the last pulse trigger has expired, the output will be 0. Always remember that you need to multiply the output of the process function to obtain a Eurorack-compatible trigger (e.g. if you want to generate a 10 V trigger, you need to multiply by 10.f).

Figure 6.5 shows a trigger pulse generated by PulseGenerator, where the trigger method is first called telling the impulse duration. At each process step, the output is generated, checking whether the trigger duration time has passed, and thus yielding the related output value.

For modules generating a clock signal, a good duration value is 1 ms (i.e. 0.001 s, or 1e-3f in the engineering notation supported by C++). A positive voltage of 10 V is OK for most uses.

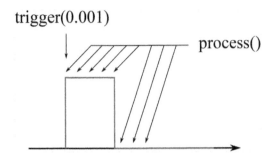

Figure 6.5: Schematic functioning of the PulseGenerator object.

We now move to the module implementation, discussing how to convert bpm to a period expressed in Hz, and how the floating-point counter algorithm described above is put into practice.

The clock module is based on the struct AClock, a Module subclass, shown below:

```
struct AClock : Module {
  enum ParamIds {
        BPM_KNOB,
        NUM_PARAMS,
  };
  enum InputIds {
        NUM_INPUTS,
  };
  enum OutputIds {
        PULSE_OUT,
        NUM_OUTPUTS,
  };

  enum LightsIds {
        PULSE_LIGHT,
        NUM_LIGHTS,
  };

  dsp::PulseGenerator pgen;
  float counter, period;

  AClock() {
        config(NUM_PARAMS, NUM_INPUTS, NUM_OUTPUTS, NUM_LIGHTS);
        configParam(BPM_KNOB, 30.0, 360.0, 120.0, "Tempo", " BPM");
        counter = period = 0.f;
  }

  void process(const ProcessArgs &args) override;
};
```

The enums define a knob for bpm selection, one output, and a light to show when the trigger is sent. The knob is configured with minimum, maximum, default values, and with two strings: the name of the parameter and the unit, in this case beats per minute (bpm). One spacing character is intentionally inserted to the left of bpm. The module also contains a PulseGenerator object and two auxiliary integer variables, initialized to 0.

The process function is pretty short, thanks to the use of the PulseGenerator object:

```
void AClock::process(const ProcessArgs &args) {

    float BPM = params[BPM_KNOB].getValue();
    period = 60.f * args.sampleRate/BPM; // samples

    if (counter > period) {
        pgen.trigger(TRIG_TIME);
        counter -= period; // keep the fractional part
    }

    counter++;
    float out = pgen.process( args.sampleTime );
    outputs[PULSE_OUT].setVoltage(10.f * out);
    lights[PULSE_LIGHT].setBrightness(out);
}
```

In the process function, we first convert the bpm value to a period expressed in Hz, which tells us how distant the triggers are far from each other in time. A bpm value expresses how many beats are to be sent in a minute. For the sake of simplicity, our clock sends one trigger pulse for each beat. To get the number of pulses per second, we need to divide the bpm by 60 seconds. For instance, a bpm of 120 is equivalent to a frequency of 2 Hz (i.e. two beats per second). However, we need to convert a frequency to a period. By definition, a frequency is the reciprocal of a period. Remember that the reciprocal of a division is a division with inverted order of the terms (i.e. $(BPM/60)^{-1} = 60/BPM$). Finally, the period is in seconds, but we are working in a discrete-time setting, where the sampling period is the basic time step. We thus need to multiply the seconds by the number of time steps in a second to get the total time steps to wait between each trigger. As an example, if the bpm is 120, the period is 0.5 s, corresponding to 22,050 time steps at a sampling rate of 44,100. This means that each 22,050 calls to the process function, we need to trigger a new pulse.

Back to the code – once the period is computed, we compare it against the counter, increasing by one unity at each time step. When the counter is larger than the period, a trigger is started by calling the trigger method of the PulseGenerator pgen. At this point, if we would reset the counter to 0, we would keep cumulating the error given by the difference between period and counter at each pulse. Instead, removing the period from counter (which is always greater or equal than period when we hit the code inside the if) enables compensation for this error, as discussed in the opening of this section.

At the end of the process function, we increase the counter and we connect the output of the pulse generator to the output of the module.

As promised, we will now discuss the use of lights and show how to obtain smooth transitions. The first way to turn the light on while the pulse trigger is high is to simply set its value, as we did in the previous module, using the setBrightness method.

There is one issue with this code, however. The pulse triggers are so short that we cannot ensure that the light will be turned on and be visible to the user. If the video frame rate is approximately 60 Hz (i.e. the module GUI is updated each 16 ms), it may happen that the pulse turns on and off in the time between two consecutive GUI updates, and thus we will not see its status change.

To overcome this issue, there a couple of options:

- Instantiate a second pulse generator with trigger duration longer than 16 ms that is triggered simultaneously to pgen but lasts longer for giving a visual cue.
- Use another method of the Light object, called `setSmoothBrightness`, which rises instantly but smooths its falling envelope, making it last longer.

The first option is rather straightforward but inconvenient. Let us discuss the second one. The `setSmoothBrightness` method provides an immediate rise of the brightness in response to a rising voltage and a slow fall in response to a falling voltage, improving persistence of the visual cue.

To invoke the setSmoothBrightness method to follow the pulse output, you can do as follows:

```
lights[PULSE_LIGHT].setSmoothBrightness(out, 5e-6f);
```

where the first argument is the brightness value and the second argument scales the fall duration. The provided value gives us a nice-looking falling light.

Well done! If you are still with me, you can play with the code and follow on with the exercises. This chapter has illustrated how to move from an algorithm to its C++ implementation. From now on, most modules will implement digital signal processing algorithms of increasing complexity, but don't worry – we are going to do this step by step.

Exercise 1

Try implementing a selector knob to choose whether the bpm value means beats per minute or bars per minute. The selector is implemented using an SvgSwitch subclass.

Exercise 2

A useful addition to the module may be a selector for the note duration that divides or multiplies the main tempo. Try implementing a selector knob to choose between three different note durations:

- *Half note.* The clock runs at half speed.
- *Quarter note.* Four beats in a bar, as usual.
- *Eighth note.* The clock runs two times faster.

You can implement this switch using the three-state SvgSwitch.

6.6 Sequencer Module

TIP: In this section, you will learn about an important Rack object, the SchmittTrigger. This allows you to respond to trigger events coming from other modules.

An N-step sequencer stores several values, ordered from 1 to N, and cycles through them at each clock event. Values may be continuous values (e.g. a voltage to drive an oscillator pitch in order to create a melodic riff) or binary values to tell whether at the nth step a drum should be triggered or not. Of course, in modular synthesis, creative uses are encouraged, but to keep it simple at this time we will just think of a step sequencer as a voltage generator that stores N values and outputs one of them sequentially, with timing given by a clock generator.

A step sequencer is a simple *state machine* that transitions from the first to the last state and resets automatically at the last one. For each state, the module sends an output value that is set using a knob. The simplest interface has one knob per state, allowing the user to change any of the N values at any time.

In this section, we want to build a basic eight-step sequencer, with eight knobs and a clock input (see Figure 6.6). At each pulse of the clock, the state machine will progress to the next state and start sending out a fixed voltage equal to the one of the knob corresponding to the current state. Eight lights will be added to show which one is the currently active status.

The challenge offered by this module is the detection of trigger pulses on the input. This can be simply done using the SchmittTrigger object provided by Rack. A Schmitt trigger is an electronic device that acts as a comparator with hysteresis. It detects a rising edge and consequently outputs a high value, but avoids turning low again if the input drops below the threshold for a short amount of time. The SchmittTrigger implemented in Rack loosely follows the same idea, turning high when the input surpasses a fixed threshold and turning low again only when the input drops to zero or below. It provides a process(float in) method that takes a voltage value as input. The method returns true when a rising edge is detected, and false otherwise. The object relies on a small state machine

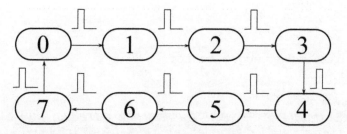

Figure 6.6: An eight-step sequencer seen as a state machine. Transitions between states are unidirectional and are issued by an input trigger. For each state, a sequencer module outputs the voltage value corresponding to the current state. Conventional sequencers have one knob per state, allowing for easy selection of the values to be sent to output. Please note that an additional reset button may allow transitions to happen from the current state toward the first one.

that also detects the falling edge so that the user can always check whether the input signal is still in its active state or has already fallen down, using the isHigh method.

Our sequencer module, ASequencer, will react to each rising edge at the clock input and advance by one step. The Module subclass follows:

```cpp
Struct Asequencer : Module {
  enum ParamIds {
        PARAM_STEP_1,
        PARAM_STEP_2,
        PARAM_STEP_3,
        PARAM_STEP_4,
        PARAM_STEP_5,
        PARAM_STEP_6,
        PARAM_STEP_7,
        PARAM_STEP_8,
        NUM_PARAMS,
  };
  enum InputIds {
        MAIN_IN,
        NUM_INPUTS,
  };
  enum OutputIds {
        MAIN_OUT,
        NUM_OUTPUTS,
  };

  enum LightsIds {
        LIGHT_STEP_1,
        LIGHT_STEP_2,
        LIGHT_STEP_3,
        LIGHT_STEP_4,
        LIGHT_STEP_5,
        LIGHT_STEP_6,
        LIGHT_STEP_7,
        LIGHT_STEP_8,
        NUM_LIGHTS,
  };


  dsp::SchmittTrigger edgeDetector;
  int stepNr = 0;

  Asequencer() {
        config(NUM_PARAMS, NUM_INPUTS, NUM_OUTPUTS, NUM_LIGHTS);
        for (int i = 0; i < Asequencer::NUM_LIGHTS; i++) {
             configParam(PARAM_STEP_1+i, 0.0, 5.0, 1.0);
        }


  }
```

```
void process(const ProcessArgs &args) override;

};
```

As you can see, the knobs also provide storage of the state values.

The process function is as simple as follows:

```
void Asequencer::process(const ProcessArgs &args) {

    if (edgeDetector.process(inputs[MAIN_IN].getVoltage())) {
        stepNr = stepNr % 8; // avoids modulus operator
    }

    for (int l = 0; l < NUM_LIGHTS; l++) {
        lights[l].setSmoothBrightness(l == stepNr, 5e-6f);
    }

    outputs[MAIN_OUT].setVoltage(params[stepNr].getValue());
}
```

The SchmittTrigger processes the input signal in order to reveal any trigger pulse. Whenever a rising edge is detected, the current step is increased, with the caveat that we need to reset it to 0 when we reach 7. We can do that using the C++ modulus operator "%" as shown above, or the bit manipulation trick shown in Section 2.13. You will find the latter in the actual implementation of the ABC plugins.

Exercise 1

Now that we have both a Clock and a Sequencer module, we can build musical sequences by driving the sequencer with a clock at a given bpm. Try building a simple musical phrase using the sequencer, the clock, and a VCO.

Exercise 2

It may be useful to reset the step counter at any time using a button or a gate input. Try implementing both using a SchmittTrigger. It will help to debounce the button and detect rising edges on the input signal.

Exercise 3

Try implementing a binary step sequencer, replacing knobs with switches, for programming drums. You can use the CKSS parameter type. It has two states, suiting your needs.

6.7 Binary Frequency Divider

TIP: In this section, you will learn how to encapsulate and reuse code.

Frequency-dividing circuits and algorithms are necessary to slow down a clocking signal (i.e. a periodic pulse train) to obtain different periodic clock signals. In digital electronics (and in music production too!), these are usually called clock dividers. Many usage examples of frequency dividers exist. Back in the analog era, DIN SYNC clocks had no standard, and to synchronize a machine employing 48 pulses per quarter note (ppq) with a slave machine requiring 24 ppq, one needed to use a clock divider with a division factor of 2. In modular synthesis, we can use a clock divider for many purposes, including sequencing, stochastic event generation, and MIDI synchronization (remember that MIDI clock events are sent with a rate of 24 ppq). Binary frequency-dividing trees are employed to obtain eighth notes, quarter notes, half notes, and so on from a fast-paced clock signal. For this reason, we may want to implement a tree of binary frequency dividers, which allows us to get multiple outputs from one single clock, with division factors of 2, 4, 8, 16, and so on. We may see this as an iterative process, where output[i] has half the frequency of output[i-1] (i.e. the same as implementing multiple factor-2 frequency dividers and cascading each one of them).

There are many C++ implementations out there of binary frequency-dividing trees even in the VCV Rack community. Some of them are based on if/else conditional statements. Here, we want to emphasize coding reuse to speed up the development. We will implement a clock divider by creating a struct implementing a factor 2 divider and cascading multiple instances of this, taking each output out.

We start with the implementation of a 2-divider object implemented as follows:

```cpp
struct div2 {
  bool status;

  div2() { status = false; }

  bool process() {
        status ^= 1;
        return status;

  }
};
```

To divide the clock frequency by a factor of 2, we can simply output a high voltage every second pulse we receive as input. In C++ terms, we may toggle a 25-Boolean variable status every two calls of its process() function. This is done by XOR-ing the status with 1.

The module, Adivider, has a bunch of div2 objects, one per output. Each div2 object is connected to a PulseGenerator, which in turn sends its signal to the output. Each divider feeds another divider, except for the last one, so that at each stage the toggling frequency slows down.

We also use the SchmittTrigger to detect edges in the input signal. In this case, we will activate at the rising edge of an incoming clock signal, initiating the process of going through our clock dividers.

This is done by a method that must be called recursively to explore the cascade of our div2 objects until we reach the last of our clock dividers or until the current divider is not active. In other words, if the current divider activates, it propagates the activation to the next one. The function is called iterActiv because it iterates recursively looking for activations.

The Module subclass is shown below:

```
struct ADivider : Module {
  enum ParamIds {
      NUM_PARAMS,
  };
  enum InputIds {
      MAIN_IN,
      NUM_INPUTS,
  };
  enum OutputIds {
      OUTPUT1, // this output will be hidden
      OUTPUT2,
      OUTPUT4,
      OUTPUT8,
      OUTPUT16,
      OUTPUT32,
      NUM_OUTPUTS,
  };

  enum LightsIds {
      LIGHTS1,
      LIGHT2,
      LIGHT4,
      LIGHT8,
      LIGHT16,
      LIGHT32,
      NUM_LIGHTS,
  };

  div2 dividers[NUM_OUTPUTS];
  dsp::PulseGenerator pgen[NUM_OUTPUTS];

  void iterActiv(int idx) {
      if (idx > NUM_OUTPUTS-1) return; // stop iteration
      bool activation = dividers[idx].process();
      pgen[idx].trigger(TRIG_TIME);
      if (activation) {
          iterActiv(idx+1);
```

```
        }
    }

    dsp::SchmittTrigger edgeDetector;

    ADivider() {
        config(NUM_PARAMS, NUM_INPUTS, NUM_OUTPUTS, NUM_LIGHTS);
    }

    void process(const ProcessArgs &args) override;
};
```

Finally, we implement the processing with the cascade of div2 objects. Please note that we start the recursive iteration only when the SchmittTrigger activates on a clock edge. Each output is processed so that it gives a +10 V pulse if the corresponding PulseGenerator is high:

```
void ADivider::process(const ProcessArgs &args) {

    if (edgeDetector.process(inputs[MAIN_IN].getVoltage())) {
        iterActiv(0); // this will run the first divider (/2) and
            iterate through the next if necessary
    }

    for (int o = 0; o < NUM_OUTPUTS; o++) {
        float out = pgen[o].process( args.sampleTime );
        outputs[o].setVoltage(10.f * out);
        lights[o].setSmoothBrightness(out, 0.5f);
    }

}
```

Before closing this section, we may want to note that the execution time of this implementation is quite variable as in some cycles there will be no call to iterActiv, while in some other cases this will be called recursively until the last output. For this module, we are not worried by this because the number of iterations is bounded by NUM_OUTPUTS and the function is quick to execute, but in DSP practice recursion is possibly dangerous. Always weigh the pros and cons and evaluate your code thoroughly. In your DSP development, take into account that one overly slow processing cycle may create a domino effect on other DSP modules or system functions, with unpredictable results. Suppose that you can implement the same algorithm in two ways: one with constant computational cost and another with variable computational cost. If the latter has a lower computational cost, on average, but with rare peaks of computing requirements, the first solution may still be better: the system will be more robust to unpredictable computing peaks. Of course, it depends a lot on the application, and experience will teach what is best.

A final remark about the panel graphics. Take a look at the background SVG that was created for ADivider. The use of a vertical spacing of 40 px was easy to accomplish in Inkscape, since the Shift+Down combination shifts the selected object by 20 px. Also, it is very easy to create vertical lines of 40 px length thanks to the ruler function in the status bar of Inkscape. This tells the angle and distance (i.e. the length) of your line.

Exercise 1

What if you want to reset all your counters (e.g. for a tempo change or to align to a different metronome)? You need a reset method in div2 to be called for each divider at once. You also need a button, and you need this button to be processed only at its pressure, using a SchmittTrigger. Your turn!

Exercise 2

There are a few lines of code we did not explore - those related to the widget. There is nothing new in these lines; however, I want you to take a small challenge now: since the number of outputs and lights is bounded by NUM_OUTPUTS, you may try to write the widget code with a for loop that instantiates a light and an output port for each of the items in the OutputIds enum. Of course, each port and light will have coordinates that are dependent on the iteration counter. Remember to connect each port and light to the correct item in the respective enum.

Exercise 3

We ended up coding a few lines to get a nice and useful module. Compare this module with other modules in the VCV Rack community performing the same function and look at their code. Try to understand how they work and make a critical comparison trying to understand in what regards this module is better and what could represent a disadvantage. Take into account computational cost considerations, reusability, code readability, memory allocation, and so on.

Exercise 4

The dsp namespace (see include/dsp/digital.hpp) has a ClockDivider struct. Try to create a clock divider tree module using this struct and compare it to the ADivider module.

6.8 Random Module

 TIP: In this section, you will learn how to generate random numbers using Rack utility functions.

Random number generation is an enabling feature for a large array of compositional techniques, allowing unpredictable behaviors to jump in or even leaving the whole generation process to chance. Moreover, generating random numbers at sampling rate is equivalent to generating noise, and thus you can generate random numbers at sampling rate to create a noise generator. Finally, random numbers are used to randomize the parameters of your module or can be necessary for some numerical algorithms (e.g. to initialize a variable). For a recap on random signals and random generation algorithms, please refer to Section 2.12.

VCV Rack provides random number generation routines in its APIs, under the random namespace. These are:

- `void random::init()`. Provides initialization of the random seed.
- `uint32_t random::u32()`. Generates random uniform numbers in the range of a 32-bit unsigned int variable.
- `uint64_t random::u64()`. Generates random uniform numbers in the range of a 64-bit unsigned int variable.
- `float random::uniform()`. Generates random uniform float numbers ranged 0 to +1.
- `float random::normal()`. Generates random float numbers following a normal distribution with 0 mean and standard deviation 1.

Please note that only the normal distribution has zero mean, and thus the other three will have a bias that is half the range. Of course, from these APIs provided by Rack, you can obtain variations of the normal and uniform distributions by processing their outputs. By adding a constant term, you can alter the bias. By multiplying the distributions, you change the range and, accordingly, the variance. Furthermore, by squaring or calculating the absolute value of these values, you get new distributions.

In this section, we describe how a random number generator module, ARandom, is developed to generate random signals at variable rate. Our module will thus have one knob only, to set the hold time (i.e. how much time to wait before producing a new output value). This is not unlike applying a zero-order hold circuit to a noise source (i.e. sampling and holding a random value for a given amount of time). However, equivalently, and conveniently, we can just call the random generation routine when we need it. The knob will produce, on one end, one value for each sampling interval, producing pure noise, or one value per second, producing a slow but abruptly changing control voltage. An additional control that we can host on the module is a switch to select between the two built-in noise distributions (normal and uniform).

The module struct follows:

```
struct ARandom : Module {
  enum ParamIds {
      PARAM_HOLD,
      PARAM_DISTRIB,
      NUM_PARAMS,
  };
  enum InputIds {
      NUM_INPUTS,
  };
  enum OutputIds {
      RANDOM_OUT,
      NUM_OUTPUTS,
  };

  enum LightsIds {
      NUM_LIGHTS,
```

```
};

enum {
      DISTR_UNIFORM = 0,
      DISTR_NORMAL = 1,
      NUM_DISTRIBUTIONS
};

int counter;
float value;

ARandom() {
      config(NUM_PARAMS, NUM_INPUTS, NUM_OUTPUTS, NUM_LIGHTS);
      configParam(PARAM_HOLD, 0, 1, 1, "Hold time", " s");
      configParam(PARAM_DISTRIB, 0.0, 1.0, 0.0, "Distribution");
      counter = 0;
      value = 0.f;
}

void process(const ProcessArgs &args) override;
};
```

The `ParamIds` enum hosts the hold time knob and a distribution selector. We enhance code readability by adding another enum to the struct, associating the name of the two distributions to two numbers (i.e. the two values of the PARAM_DISTRIB switch). Two variables are declared, a float that stores the last generated value and an integer storing the number of samples since we last generated a random number. The latter will increase once per sample time, and be compared to the value to see if it is time to generate a new sample.

Before going into the details of the process function, let us consider the configParam for the hold time knob. The configured range goes from 0 to 1 second, for simplicity. Obviously, when the knob is turned to the minimum value, the hold time will be equivalent to one sample, for continuous generation of random values. On the other hand, the hold time will be equivalent to F_s samples.

We are getting now to the process function. This only has to read the parameters and let the random routines generate a new value, if it is time to do that. The code follows:

```
void ARandom::process(const ProcessArgs &args) {
  int hold = std::floor(params[PARAM_HOLD].getValue() * args.
    sampleRate);
  int distr = std::round(params[PARAM_DISTRIB].getValue());

  if (counter ≥ hold) {
      if (distr == DISTR_UNIFORM)
            value = 10.f * random::uniform();
      else
            value = clamp(5.f * random::normal(), -10.f, 10.f);
      counter = 0;
  }
```

```
    counter++;

    outputs[RANDOM_OUT].setVoltage(value);
  }
```

The routine first reads the status of the parameters. The number of samples to wait until the next number is generated is stored into the variable hold and is computed by multiplying the value of the knob by the sampling rate. The PARAM_HOLD knob goes from 0 to 1 seconds.

If the counter is equal to[3] or greater than hold, the module generates a number according to the chosen distribution. When hold is 0 or 1, the module generates a new sample for each call of process.

Each distribution is scaled properly to fit the Eurorack voltage range: random uniform will be ranged 0 to 10 V, random normal will give random numbers around 0 V, possibly exceeding the −10 to +10 V range, and thus clamping it to the standard voltage range is a good idea.

The rest of the module is quite standard, so you can refer to the source code to see how the module is created.

Exercise 1

The hold knob maps time linearly in the range [0, 1]. The first quarter turn of the knob compresses most of the useful values. What about replacing time with tempo? By considering the value on a beats per minute (bpm) scale, we have a musically meaningful distribution of time values along the knob excursion range that provides finer resolution where it matters more. Specifically, the knob should display bpm values in the range 15 to 960. This can be done according to the following:

$$BPM = 60 \cdot 2^v \tag{6.2}$$

where the knob value is $v \in [-2, 4]$. This is easily achieved exploiting optional arguments of the Module::configParam method.

The process method should compute the values according to the above, and convert the bpm value into a time in samples according to the following:

$$hold|_{\text{smp}} = \frac{60 \cdot F_s}{BPM} = \frac{F_s}{2^v} \tag{6.3}$$

Such a module may be useful for generating random values at a steady tempo, but it is not generating noise anymore. One workaround to this would be to increase the upper bpm limit to a value that makes the hold time equal or shorter than one sample. Do consider, however, that the sample duration depends on the sampling rate, and thus one would have to adjust the upper limit if the engine sample rate is changed by the user. Unfortunately, Rack v1 APIs do not allow for this, and thus your best solution is to consider the worst case (highest sampling rate).

One last tip related to this example: change the strings in configParam to indicate that the knob is a tempo, not an absolute time.

Exercise 2

The module of Exercise 1 is synchronized to an internal bpm tempo. This is useful for generating rhythmic random modulations, but cannot synchronize with other clock sources. Modify the ARandom module to include an input port that allows synchronization to an external clock source. By checking whether the input port is connected, you can exclude the internal clock generation mechanism. This system is the equivalent of a sample and hold circuit applied to a random generator.

6.9 To Recap: Exercise Patches

Check the online resources to see how to complete these exercises.

6.9.1 Creating a Bernoulli Distribution and Bernoulli Gate

A Bernoulli distribution is a random distribution that only allows two values, 0 and 1. You can create that by using the ARandom module and the AComparator. The random numbers need to be compared against a threshold value (e.g. using LOG instrument "Precise DC Gen"). The higher the threshold, the lower the number of "1" outputs.

A Bernoulli gate is a binary multiplexer that sends the input to either of the two outputs. You can modify the AmuxDemux to let a CV in to select which output should be used and send the Bernoulli random values generated using AComparator to control the output port.

6.9.2 Polyrhythms

By using two sequencers driven by the same clock, you can generate interesting rhythmic patterns. We can connect AClock to two ASequencer modules, but they both would have eight steps for a 4/4 time signature. You can use ADivider to have them running at multiple rates. However, if you want to have one running with a 4/4 signature and the other one at a 3/4 signature, for example, you need to modify ASequencer to add a knob to select how many steps to loop (from one to eight), or have a Reset input and ensure it is triggered at the right time (after 3/4).

Notes

1 Please note, we are not going to place screws in the ABC modules for the sake of simplicity.
2 The period of a year consists of 365.24 days approximately.
3 *Tip*: In principle, the condition to trigger the generation of a new number should be `counter == hold`. However, this comparison may produce bugs if for some reason (e.g. an overrun due to preemption to Rack or a variable overwritten by accident by a buggy write access) we skip this code. From the next time on, the condition will be never verified, because counter will always be larger than hold and the algorithm will fail in generating a new random number. Thus, it is always better to use the equal to or greater than condition, to avoid such issues.

CHAPTER 7

Getting Serious: DSP "Classroom" Modules

By now, you have the basics to develop Rack modules. In this chapter, we will start delving into DSP stuff, which is ultimately what we are aiming for when we program in VCV. This chapter will guide you into several examples and will report, along the way, useful tips for DSP programming. Finally, one of the modules will be modified to allow for polyphony.

The modules presented in this chapter are shown in Figure 7.1.

7.1 Handling Polyphony Properly

Before starting with audio modules, let us review some basics related to ports and cables.

Up to now, we have seen two methods, Port::setVoltage and Port::getVoltage, for writing and reading a value to/from a port, respectively. This is OK for mono CV ports; however, a more proper management of the ports requires a bit more explanation.

The first issue we need to address is the presence of polyphonic modules. These are modules that have polyphonic output ports. How should we handle polyphonic cables connected to a mono input? Recalling Section 3.6.5, the general rule that all developers should follow is to discard exceeding channels if the input expects a CV, or sum all the channels if the input expects an audio signal. These rules minimize user dissatisfaction. Think, for example, of a mono filter receiving multiple audio signals from a poly oscillator. By summing all signals carried by the polyphonic cable, the filter will not discard any of the notes coming from the oscillator. It will not handle them separately, but at least it will not kill any of them. Let us now consider a polyphonic CV cable connected to the cutoff frequency input of the same mono filter. In this case, taking the first channel is the best way to get at least one usable CV to drive the cutoff frequency, as the sum of CV signals would make no sense for the control of this parameter.

The modules designed in the previous chapter are all handling CV or hybrid CV signals, and thus the use of the getVoltage method is justified. In this section, when developing audio modules, such as the state-variable filter, we shall use the following method:

```
float getVoltageSum()
```

This sums together all channels. It is advisable to scale the result by the number of active channels to avoid overly large signals.

Figure 7.1: The modules presented in this chapter.

For what concerns polyphonic outputs, we shall see in Section 7.6 how to turn a monophonic module into a polyphonic one with a few minor tweaks. The setVoltage method, by default, writes the value to channel 0 of the polyphonic cable. However, the second argument can be exploited to write the value to any of the 16 channels:

void setVoltage(float voltage, **int** channel = 0)

As an example, the following two lines write to channels 0 and 1, respectively:

```
outputs[<outputID>].setVoltage(<value>);
outputs[<outputID>].setVoltage(<value>,1);
```

The system should also be informed about the number of channels that the cable is carrying, otherwise it will treat it as a default mono cable, even though nonzero values have been written to the upper channels. The method for setting the number of channels is:

```
void setChannels(int channels)
```

This method also takes care of zeroing those channels that are not used.

When designing polyphonic modules, special care must be taken to match the cardinality of the input and output channels. Let us consider a poly filter with N inputs. This should produce N filtered outputs by instantiating N filter engines. Similarly, a poly oscillator with N V/OCT inputs should produce N outputs. How about the secondary inputs?

Consider, for example, a filter with N inputs and M cutoff CV:

- If $M = 1$, the cutoff CV should be copied to all instances of the filter engine.
- If $M \geq N$, the first N cutoff CVs should be used, one for each instance of the filter engine; the rest is discarded.
- If $1 < M < N$, the available cutoff CVs should be used for the first M instances of the filter engine; the rest of the engines will get 0 V.

This is allowed by the following method:

```
float getPolyVoltage(int channel)
```

The method handles all cases, provided that 0 V is written to input channels that are not used (this is normally done by the setChannels method, executed beforehand by the module providing the cable).

All this will be discussed with an example later in this chapter.

7.2 Linear Envelope Generator

Envelope generators (EGs) are fundamental to shape the dynamic of a sound as well as the rhythm. Most envelope generation schemes follow the so-called ADSR scheme, as described in Section 1.3. The easiest way to generate envelopes in the digital domain is by creating linear ramps. This is done by cumulating small steps, evaluated in order to obtain a desired ramp time. The step determines the rise or fall time: the bigger the step, the faster the ramp. The ramp grows if the step is positive, or it falls if it is negative.

Linear ramps are not the best in psychoacoustic terms, since they do not take into account the logarithmic perception of the signal amplitude, but they provide a nice starting point, and they will be of use for many purposes in a modular rack.

The module is based on a gate input to determine the stage of the envelope: the attack starts at the rising edge, while the release starts at the falling edge. The four ADSR values are set by knobs.

The functionalities to embed into the process function are:

- evaluating the step (i.e. the rate of increase/decrease for each stage, according to knob values);
- computing the state machine to advance through the stages; and
- computing the next envelope value by summing the step to the previous output.

The evaluation of the ramp values is pretty straightforward. Try figuring it out yourself. Assuming we are starting from 0 and we want to reach the target value T_g in A_t seconds, at a sampling rate of F_s, how many samples do we need in order to reach the target value?

The answer is: the target value is reached in $A_t \cdot F_s$ samples. The step size is thus the target value divided by the number of steps to reach it, and thus the resulting formula will be:

$$A_{step} = \frac{T_g}{F_s \cdot A_t + \epsilon} \tag{7.1}$$

where ϵ is a very small value to avoid dividing by zero when A_t is zero.

If this is clear, we can generalize the reasoning to the other two step evaluations (i.e. that of the decay D_{step} and the release R_{step}). In these cases, the target value is the distance to travel from the top (1.0) to the sustain level and from the sustain level to zero. To compute Rstep, we also add the to the numerator, to avoid zero when the sustain is zero.

This is the code for computing the steps:

```
float sus = params[PARAM_SUS].getValue();
float Astep = 1.f/(EPSILON + args.sampleRate * params[PARAM_ATK].
getValue());
   float Dstep = (sus - 1.0)/(EPSILON + args.sampleRate * params[PARAM_
DEC].getValue());
   float Rstep = -(sus + EPSILON)/(EPSILON + args.sampleRate *
params[PARAM_REL].getValue());

Astep = clamp(Astep, EPSILON, 0.5);
Dstep = std::max(Dstep, -0.5f);
Rstep = std::max(Rstep, -1.f);
```

Please note that the step constants are clamped to avoid overshoots, because the smallest attack times may yield a step larger than the entire excursion range, depending on the EPSILON and the sampling rate.

A state machine determines the envelope generation stage and requires a SchmittTrigger to detect a rising edge on the gate input for determining the attack phase. If the state machine is properly built, the release phase is inferred by a low value of the gate signal. Overall, the finite-state machine corresponding to an ADSR EG is reported in Figure 7.2.

The state machine we are building needs to go to Attack any time the rising edge is detected. While the gate input is active, the state machine controls the status of the envelope, conditionally switching

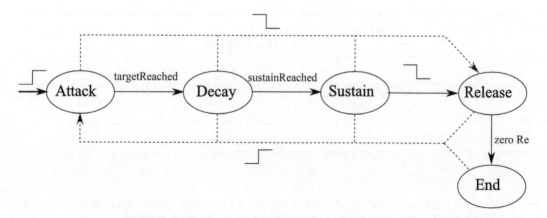

Figure 7.2: The state machine corresponding to an ADSR envelope generator. Jumps to the Attack phase or to the Release phase on gate rising or falling edge are shown by the dashed line.

to the next phase: when the target value is reached, the Attack ends and the Decay starts. Similarly, when the sustain value is reached, the Decay stage ends, and the Sustain stage begins. In Release, when zero is reached, the envelope generation is considered closed and zero will be written to the output until a new rising edge gate input is detected.

The code follows:

```
bool gate = inputs[IN_GATE].value ≥ 1.0;
if (gateDetect.process(gate)) {
        isAtk = true;
        isRunning = true;
}

if (isRunning) {
        if (gate) {
                //Attack
                if (isAtk) {
                        env += Astep;
                        if (env ≥ 1.0)
                                isAtk = false;
                }
                else {
                        //Decay
                        if (env ≤ sus + 0.001) {
                                env = sus;
                        }
                        else {
                                env += Dstep;
                        }
                }
        } else {
                //Release
                env += Rstep;
                if (env ≤ Rstep)
                        isRunning = false;
        }
} else {
        env = 0.0;
}

if (outputs[OUT_ENVELOPE].isConnected()) {
        outputs[OUT_ENVELOPE].setVoltage(10.f * env);
}
```

One thing worth mentioning: the target value is set to 1.0 for simplicity, and the generated envelope is multiplied by 10 in the last lines to comply with the 10 V Eurorack standard.

Exercise 1

In Equation 7.1, we used a small value to avoid division by zero. Conversely, we can allow $A_t \geq \epsilon$ when we set the knob excursion range in the configParam method and save the sum in the denominator of Equation 7.1. Try this out in your code for all the step evaluations and the time input values. You will not save much in computational terms here, but, in the code optimization stage, reducing the number of operations that are done at least once per sample sums up to an improved outcome.

Exercise 2

Generalizing the concept of the EG, you can devise a module that generates an envelope signal by ramping between N points, of which you can set the time and the amplitude by knobs. Figure 7.3 provides an idea of this envelope generation method. Please note, this module also accepts a trigger signal as input, since the falling edge of the gate has no use.

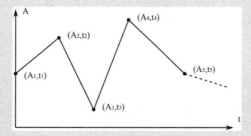

Figure 7.3: An envelope generator following a generic scheme with N points with adjustable amplitude and time. Amplitude and time are shown as coordinates of the points $t(A_i, t_i)$.

Exercise 3

Referring to the module of Exercise 2, can you allow the state machine to loop indefinitely between the N points? What kind of module are you obtaining if the loop period is very short, say lower than approximately 20 ms?

7.3 Exponential Envelope Generator

TIP: In this section, you will learn how to model a simple RC circuit, making sense of those high school lessons on electric circuits. We will take the circuit from the analog to the digital domain, encapsulate it in a class, and use it to process a signal.

The linear envelope generator we developed in the previous section has somewhat limited use in musical synthesizers for two reasons: our perception of loudness is not linear, and analog envelope generators are usually exponential, due to the simplicity of generating exponential rise and fall curves. For this reason, we are now going to implement an envelope generator with exponential curves, writing the process function from scratch. We could just adapt the code of the linear envelope generator to this task by calculating the exponential of the linear ramp, but this could be expensive. Transcendental functions, such as logarithms, exponentials, and goniometric functions from the math library, are expensive as they are computed with a high precision. In this context, we can obtain an exponential decay much more efficiently.

There are a few ways to implement an exponential ADSR. The one I prefer to report here is easy, reusable, and efficient, and it does not involve explicit evaluation of exponentials.

7.3.1 A Virtual Analog RC Filter

Suppose we don't know how to generate an exponential ramp. Simple high school physics concepts will guide us here. You may know that a circuit composed of a resistor and a capacitor (RC) has an exponential voltage change when subject to sudden input voltage changes. If, for instance, at a certain time, the capacitor is charged and it is connected to the resistor by closing a switch, the voltage across the capacitor drops exponentially, discharging through the resistor. We shall take this easy circuit as a guide to design a digital circuit that provides exponential response to input steps.

It can be shown that the voltage across the capacitor varies in time according to the following:

$$V_c(t) = V_i\left(1 - e^{-t/RC}\right) \tag{7.2}$$

In other words, it varies exponentially, decaying from an initial value V_i with a time constant that is generally defined as $\tau = RC$. Please note that τ is the time required for the signal to fall to V_i/e when the initial voltage is V_i and a null voltage is applied to the input (short circuit). The RC circuit is also known to be a passive low-pass circuit (see Figure 7.4).

Let us move through the process of obtaining a discrete-time model of the continuous-time RC circuit. We are interested in the input–output relation. We can get this from evaluating the voltage drop across the resistance. According to Ohm's law, this is proportional to the current that flows through it according to:

$$v_R(t) = Ri(t) \tag{7.3}$$

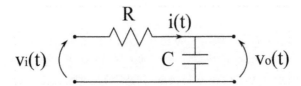

Figure 7.4: An analog first-order low-pass RC filter. If the capacitor is charged ($v_o(t) = v_i(t) = V_i$ for $t < 0$) and the input voltage drops to zero for $t = 0$, the output voltage follows a falling exponential curve.

We can also look at the circuit and note that $v_i(t) - v_o(t) = v_R(t)$. Now we can show that the current flowing into a capacitor is:

$$i(t) = C\frac{dv_o(t)}{dt} \tag{7.4}$$

which is derived from plugging the definition of capacitance $C = Q/v(t)$ into the definition of current $i(t) = dQ/dt$.

By mixing Equations 7.3 and 7.4, we see that:

$$v_i(t) - v_o(t) = RC\frac{dv_o(t)}{dt} \tag{7.5}$$

This difference equation can be discretized to obtain a discrete-time implementation of the RC filter approximation:

$$x[n] - y[n] = RC\frac{y[n] - y[n-1]}{T} \tag{7.6}$$

where we applied the definition of derivative in the discrete-time domain seen in Section 2.6 using $\Delta x = T$ to make time consistent across the two domains:

$$\frac{dv_o(t)}{dt} \rightarrow \frac{y[n] - y[n-1]}{T} \tag{7.7}$$

where T is the sampling period (i.e. $T = (1/F_s)$) and $y[n], x[n]$ are the discrete-time output and input voltage, respectively. Note that the input sequence is known and the previous output value can be stored in a memory element. We can thus rearrange Equation 7.6 to evaluate the output sequence as:

$$y[n] = ay[n-1] + (1-a)x[n] \tag{7.8}$$

where:

$$a = RC/(RC + T) \tag{7.9}$$

Equation 7.8 informs us that the output is the cumulative contribution of itself and a new input sample.

7.3.2 Implementation of the Exponential Envelope Generator

The implementation of the RC filter is very easy in code. We will implement it as a struct, in conformance to Rack objects, and we want to make it available in the form of single- and double-precision arithmetic. To do this, we use C++ templates. We thus declare the struct as a generic implementation of a typename T that will be later defined when the class is instantiated.

The class heading will thus be:

```
template <typename T>
struct RCfilter {
    ...
}
```

The processing C++ code follows:

```
T process (T xn) {
  yn = a * yn1 + (1-a) * xn;
  yn1 = yn;
  return yn;
}
```

where `yn1` stands for *y[n − 1]*.

The second line is the update step. It is necessary to prepare the current output to become the previous output at the next execution of the process function. Never exchange the order of the two lines, otherwise we are updating the element before computing.

We want to set the tau time for the integrator, and we do that by setting *a* according to Equation 7.9:

```
void setTau (T tau) {
        this->a = tau/ (tau + APP->engine->getSampleTime ());
}
```

This way, the time set from the module knobs will be the tau time (i.e. the time to reach {2/3} of the whole excursion).

Now that we have designed our RC filter, we only need the control code to build our new module. Different from the linear ADSR, this one will generate constant voltage values that the RC filter will follow. To create the attack ramp, for example, we let the RC filter follow a step signal. The RC filter will react slowly to the step signal, getting closer and closer to the constant value of the step signal, in an exponential fashion. Once the signal reaches the target value (or gets very close to it), the step function drops to the sustain level, allowing the filter to tend to its value, and so on, as shown in Figure 7.5. The gate rising and falling edges are detected, employing the SchmittTrigger class already discussed in the previous chapters.

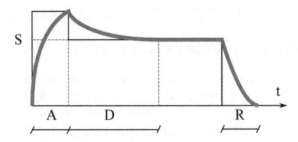

Figure 7.5: The RC filter output (thick solid line) following a step-like function (thin solid line). For its inner nature, the RC filter will decay or rise exponentially.

The RCFilter is declared using the float template:

```
SchmittTrigger gateDetect;
RCFilter<float> * rcf = new RCFilter<float>(0.999);
```

The process function is implemented below, where you can find the knob parameter reading, the gate processing, and the ADSR state machine:

```
void AExpADSR::process(const ProcessArgs &args) {
    float sus = params[PARAM_SUS].getValue();

    Atau = clamp(params[PARAM_ATK].getValue(), EPSILON, 5.0);
    Dtau = clamp(params[PARAM_DEC].getValue(), EPSILON, 5.0);
    Rtau = clamp(params[PARAM_REL].getValue(), EPSILON, 5.0);

    bool gate = inputs[IN_GATE].getVoltage() >= 1.0;
    if (gateDetect.process(gate)) {
            isAtk = true;
            isRunning = true;
    }

    if (isRunning) {
            if (gate) {
                    if (isAtk) {
                            //Attack
                            rcf->setTau(Atau);
                            env = rcf->process(1.0);
                            if (env ≥ 1.0 - 0.001) {
                                    isAtk = false;
                            }
                    }
                    else {
                            //Decay
                            rcf->setTau(Dtau);
                            if (env ≤ sus + 0.001)
                                    env = sus;
                            else
                                    env = rcf->process(sus);
                    }
            } else {
                    //Release
                    rcf->setTau(Rtau);
                    env = rcf->process(0.0);
                    if (env ≤ 0.001)
                            isRunning = false;
            }
    } else {
            env = 0.0;
```

```
        }

    if (outputs[OUT_ENVELOPE].isConnected()) {
            outputs[OUT_ENVELOPE].setVoltage(10.0 * env);
    }

}
```

Please note that the exponential function reaches the target value in a possibly infinite amount of time, and thus the condition for switching from attack to decay and from decay to sustain is referred to a threshold level, arbitrarily set to 0.001 in the code above.

All the other parts of the module follow the linear ADSR we already designed.

Note: The Rack API provides a state-variable first-order RC filter, implemented in include/dsp/filter. hpp. Its process method differs from the one we implemented here as it does not return an output value. It is so because it provides two output signals, low-pass and high-pass, that can be obtained separately after each call to the process invoking the low-pass and high-pass member functions. Furthermore, it does not have a method to set the decay rate parameter tau.

Exercise 1
This module builds from theory quite easily. We are going further in understanding bits of circuit theory and digital signal processing – well done! Now take a look at other envelope generators out there in the VCV Rack community. You will find that many implementations are less efficient compared to this one (e.g. some of them compute the exponentiation using powf(), which is not as efficient as two sums and multiplications) and they are not as simple. Of course, some of them will handle more intricate functions in addition to ADSR generation that this implementation may not allow, so try to make a fair comparison. To compare the computational resources required by each module, you can turn on the CPU Timer from the Engine menu.

7.4 Envelope Follower

 TIP: In this section, we will reuse code, showing the benefits of OOP in a DSP context.

The envelope of a signal represents its short-term peak value and outlines its extremes, as shown in Figure 7.6. This is done by extracting local peaks and smoothly interpolating between them.

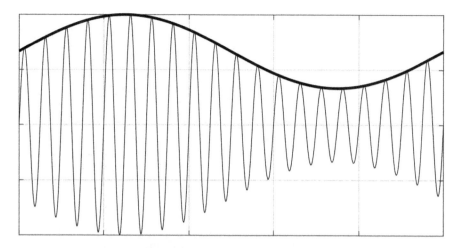

Figure 7.6: A sinusoidal signal modulated by a sine wave of lower frequency. The envelope (solid bold line) of the signal shows the modulating low-frequency sine wave.

Detecting the envelope of a signal is useful for level monitoring, dynamic effects, gain reduction, demodulation, and more. Depending on the task, we may be interested in getting the peak envelope, as in Figure 7.6, or the root mean square envelope. Here, we shall focus on the peak value.

7.4.1 Modelling a Diode Envelope Follower

How do we create an envelope follower? Historically, cheap envelope followers (or envelope detectors) were found in DSB AM radios, as discussed in Section 2.11.2. These were made employing diodes. Let us thus introduce a few notions on diodes. Diode circuits are used for a lot of applications, including clipping, envelope extraction, modulation circuits, wave rectification, and more (some of these where tackled in Section 2.11).

The diode acts similarly to a switch, producing a current when the voltage across its terminals is positive, and producing no current otherwise. One well-known model for the bipolar junction diode is the Shockley diode equation:

$$i_D(t) = I_s \left(e^{nV_T/v_D(t)} - 1 \right) \tag{7.10}$$

where $v_D(t)$ is the voltage across the diode and $i_D(t)$ is the diode current. From the equation, we observe that the time-varying current is related to the time-varying voltage, while all other terms are constants related to fabrication process, silicon properties, and temperature. Let us consider a simple circuit including a resistor, as shown in Figure 7.7. When the voltage $v_D(t)$ is positive (Figure 7.7a), the current flows across the diode, charging the capacitor and producing a voltage at the envelope detector output terminals. Differently (Figure 7.7b), the diode cannot conduct (it does produce an extremely small current $-I_s$, which is negligible in this context) and the output voltage is the voltage of the charged capacitor. A faithful model of the diode current requires some more computational power (e.g. see D'Angelo et al., 2019), which is not necessary here. In the context of this chapter,

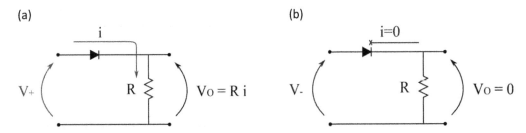

Figure 7.7: A diode-resistor circuit. When a positive voltage V^+ is applied, the output voltage is positive. When a negative voltage is applied, the output is zero and the reverse current flow is stopped by the diode.

we can very roughly approximate a diode by a switch that conducts a current when the voltage at its input is positive (current runs in forward direction) and shuts the current when the input voltage is negative (current runs in reverse direction). This is shown in Figure 7.7. Note how the diode symbol used in schematics recalls this mechanism, being composed of an arrow in the forward direction and a perpendicular line, like a barrier, in the reverse direction.

In our software applications, we can consider the diode as a box that simply applies the following C/C++ instruction:

```
y = x > 0 ? x: 0;
```

where x and y are, respectively, the input and output virtual voltage.

A simple envelope detector can be made by a series-diode and a shunt RC circuit, as depicted in Figure 7.8.

We shall thus consider two scenarios: when the diode is active, it acts as a pump that charges the capacitor, while when interdicted the circuit reduces to a RC parallel. The resistor in parallel with the capacitor will slowly discharge the capacitor, as shown in Figure 7.9b. The entity of the resistor and the capacitor imply the decay time constant. This can be left as a user parameter, as we want to keep the system usable over a wide range of input signals.

We can improve this design slightly by rectifying the negative wave too, resulting in a full-wave rectifier. This is done using a rectifying diode bridge: when four diodes are connected together as

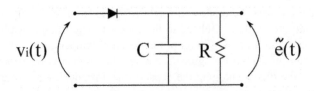

Figure 7.8: A simple diode envelope follower and the resulting output, approximating the ideal envelope.

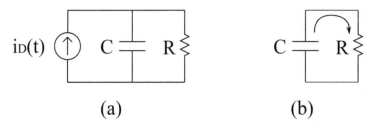

Figure 7.9: Equivalent circuit of the envelope detector during diode conduction (a) and diode interdiction (b). In the first case, the diode injects current that charges the capacitor, while during interdiction the capacitor discharges through the resistor.

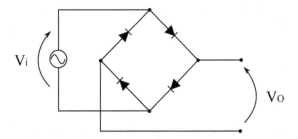

Figure 7.10: A diode rectifier or diode bridge. The output voltage takes the absolute value of the input voltage. In other words, the negative wave is rectified.

shown in Figure 7.10, the output is approximately the absolute value of the input. In C/C++, this is obtained by an if statement of the kind:

```
if (x < 0)
        y = -x;
else
        y = x;
```

or the equivalent:

```
y = (x > 0) ? x: -x;
```

In terms of efficiency, we are often repeating that conditional instructions, such as if or the "?" operator, should be avoided. However, simple instructions, such as the ones for the absolute value or for rectification, can be compiled efficient. If compiler optimizations are enabled, this is able to translate the conditional instruction to a dedicated processor instruction that implements the statement in a short number of cycles. A compiler explorer, such as https://godbolt.org/, allows you to watch how a snippet of C++ code is compiled into assembler. You can provide the pertinent compiler flags (those related to the code optimization and math libraries that the Rack makefile imposes) and observe what processor instructions are invoked to implement your C++ statements. As an example, Rack currently compiles with the following flags: -O3 -march=nocona -ffast-math -fno-finite-math-only. Try inserting the following snippet:

```
float absol(float x) {
        return (x > 0) ? x : -x;
}
```

The code is compiled into a single instruction: andps, which basically is a bitwise AND, showing how bit manipulation can do the trick. You can also watch how the standard C++ implementation of the absolute value is compiled by replacing the previous code with the following:

```
#include <cmath>

float absol(float x) {
        return std::abs(x);
}
```

The diode alone is not sufficient for recovering the envelope. We need at least an additional element, a capacitor, to store the charge when the maximum value is reached. Figure 7.11 shows how the capacitor retains the value of a rectified wave and a fully rectified wave (absolute value). The ripple in the estimated envelope must be as low as possible, and the full-wave rectifier helps in this regard. Changing the capacitor value (i.e. the discharge time) affects the ripple amount.

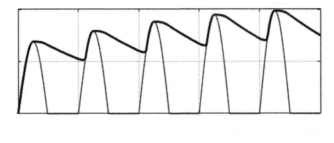

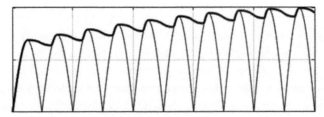

Figure 7.11: Comparison between the diode rectifier (a) and the full-wave rectifier (b). The output of the diode or the full-wave rectifier (solid thin line) is shown against the output of the envelope detector (bold solid line).

7.4.2 *Implementation of the Envelope Follower*

Starting from what we have learned in Section 7.3, we can devise an envelope follower object. The RCFilter struct was designed after a series RC circuit, which partially suits our needs: the diode-RC circuit acts as an RC filter when the diode is interdicted, as shown in Figure 7.9b, where the capacitor discharges through the resistor. However, at least one modification to RCFilter is required. The modification involves the introduction of a sort of *charge pump* method, or, out of the metaphor, a method to force the voltage across the capacitor while the input rises. This is required to simulate the fact that the voltage is forced from the input to the output when the diode is conducting.

We can derive a new struct – let us call it RCDiode – that inherits all the methods and members from RCFilter and adds the charge pump method.

```
template <typename T>
struct RCDiode: RCFilter<T> {

.../ /code goes here
};
```

The charge method sets the current output and the previous state (the stored voltage) to a given value (which is also returned for convenience):

```
T charge(T vi) {
        yn = yn1 = in;
        return vi;
}
```

Since we demanded most of the DSP code to RCDiode, the process function will result in a fairly short snippet, as follows:

```
void AEnvFollower::process(const ProcessArgs &args) {

    float tau = clamp(params[PARAM_TAU].getValue(), EPSILON, 5.0);

    rcd->setTau(tau);
    float rectified = std::abs(inputs[MAIN_IN].getVoltage());
    if (rectified > env)
            env = rcd->charge(rectified);
    else
            env = rcd->process(rectified);

    if (outputs[MAIN_OUT].isConnected()) {
            outputs[MAIN_OUT].setVoltage(lights[ENV_LIGHT].value = env);
    }

}
```

where rcd is the RCDiode object, declared as:

```
RCDiode<float> * rcd = new RCDiode<float>(0.999f);
```

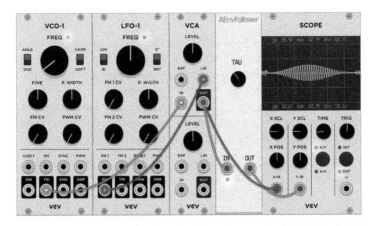

Figure 7.12: AM modulation and demodulation patch. The VCO-1 (carrier signal) is modulated by LFO-1 (low-frequency modulating signal) using a conventional VCA. The resulting signal is demodulated using the envelope follower. The scope shows the oscillating modulated signal and the envelope on top of it.

As you can see, the *charging* is done in one line as the charge method return value is assigned to env.

The tau value is clamped as it was for the exponential ADSR, to avoid overly quick or slow filter responses. When the rectified voltage is larger than the current output envelope, it takes on so that the output envelope follows the rectified voltage, and the capacitor is virtually *charged*. Otherwise, there is no input to the RC filter and the output voltage follows its natural decay.

A good way to test this module is to conduct amplitude modulation and reconstruct the envelope with this module according to the following scheme. This is shown in one of the demo patches, depicted in Figure 7.12.

Exercise 1

Try alternative calculations of the peak amplitude. You may experiment with a simple max hold algorithm with linear decay, such as:

```
float in = std::abs(inputs[MAIN_IN].getVoltage());
if (in > env + 0.001) {
        env = in;
} else {
        env += Rstep;
}
```

You need to adapt the Rstep from the linear envelope generator of Section 7.2.

Try to enumerate the differences between this and the previous method. Is the resulting envelope going to be smooth?

Exercise 2

Try adding a smoothing RC filter to the output of the envelope follower to improve the quality of the detected envelope. Can you notice the phase delay applied to the envelope signal? Does it change according to the tau parameter?

7.5 Multimode State-Variable Filter

 TIP: This section shows a fundamental building block of any musical toolchain: the digital state-variable filter. Its formulation is derived by discretization using basic virtual analog notions. Analog state-variable filters have been used in many famous synthesizers, such as the Oberheim SEM.

This section introduces you to the design and development of one of the most used digital filters in musical application. Filter design is a wide topic and research is still undergoing, with statistical and computational intelligence experts chiming in to find novel optimization techniques to fulfill filter design constraints. In musical applications, however, a few filter topologies are used most of the time and need be learned. Among these, the state-variable filter is "a must" because it has several interesting features:

• It is multimode, having low-pass, high-pass, and band-pass outputs all available at once.[1]
• It has independent control of resonance and cutoff.
• It is rather inexpensive and has good stability to sudden filter coefficients modifications.

The RC filter of Section 7.3 is nothing but a first-order low-pass filter, and it has somewhat limited applicability for processing audio signals. In this section, we are going to introduce the second-order *state-variable filter* (SVF). Its name comes from the fact that it is formulated according to a state-space model (Franklin et al., 2015), gaining advantage in several regards if compared to other filter formulations.

7.5.1 A Discrete-Time Model of the Multimode State-Variable Filter

Let us introduce the second-order SVF, with 12 dB/oct slope and high-pass, low-pass, band-pass, and notch outputs all at the same time. The differential equations of the SVF are described by the flow graph of Figure 7.13. This is the SVF in its pure mathematical form, and it can be implemented either in the analog domain as an electronic circuit or in the digital domain as an algorithm. The filter has two coefficients: the first is the cutoff frequency in radians; the second is the damping. The damping is inversely proportional to the quality factor, and thus controls the behavior of the filter around the cutoff frequency. A low damping coefficient leads to a high quality factor, with a bell-like resonance around the cutoff frequency.

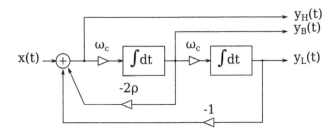

Figure 7.13: The topology of a state-variable filter. This mathematical description of the SVF can be implemented in both the analog and the digital domain.

The filter has three outputs, $y_H(t)$, $y_L(t) + y_B(t)$, the high-pass, low-pass, and band-pass outputs, respectively. Furthermore, a notch output can be calculated as $y_N(t) = y_L(t) + y_H(t)$.

The filter requires two integrators. In the analog domain, these are implemented by operational amplifiers with a feedback capacitor. In the digital domain, we can approximate an ideal integrator, as discussed in Section 2.6. With this filter topology, the choice of the integrator type is imposed by problems of delay-free loops. Indeed, if choosing a backward rectangular integrator for the first integrator, we would have the flow graph shown in Figure 7.14a. The output of the block would not be computable in a discrete-time setting as $y_B[n]$ depends on itself. Looking at the flow graph, you can extract the expression:

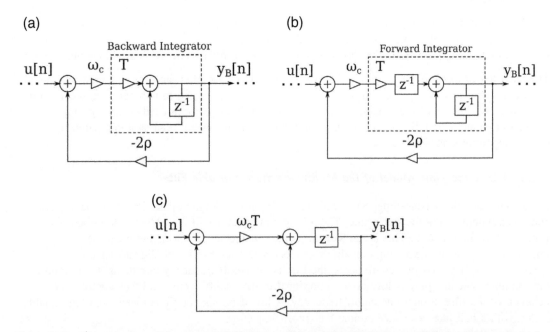

Figure 7.14: Discretization of the state-variable filter topology. The circuit surrounding the first integrator is discretized using the backward rectangular integration (a), resulting in a delay-free loop that prevents computability. If forward rectangular integration is used, the circuit is computable in a discrete-time setting (b). A more efficient version of this circuit is shown in (c), where the two delays are refactored in one.

$$y_B[n] = \omega_c T(u[n] + y_B[n]) + y_B[n-1].$$ (7.11)

This mathematical model of the filter is legitimate, but cannot be computed, and thus it would be of no use to us. By using a forward rectangular integrator, we obtain a computable, as shown in Figure 7.14b, leading to the following expression:

$$
\begin{aligned}
y_B[n] &= D\{\omega_c T(u[n] + y_B[n])\} + y_B[n-1] \\
&= \omega_c T(u[n-1] + y_B[n-1]) + y_B[n-1]
\end{aligned}
$$ (7.12)

where we denoted the delay operator as $D\{\cdot\}$ (corresponding to the z^{-1} symbol in the figures). Before moving on, there is a small step we can take to improve the efficiency of the system by removing one memory element. Since all the addends in Equation 7.12 are delayed by one sample (i.e. all are referred to time instant $[n-1]$), we can just use one delay element at the end of the forward branch that applies to all the addends, as shown in Figure 7.14c. The change in the flow graph is reflected in the code implementation. Please note that the mathematic expression is still the one of Equation 7.12, and thus we have not changed the system; we just have changed its implementation to make it feasible.

The discretization of the second integrator is not subject to computability issues now, and thus we can simply use the backward rectangular integrator that only requires one memory element. The discrete-time flow graph for the SVF is shown in Figure 7.15, where $\gamma = 2\rho$. This topology is also known as the Chamberlin filter (Chamberlin, 1985). Similar topologies have been derived by Zölzer and Wise (Wise, 2006). A different route has been taken by Zavalishin, using trapezoidal integration (not covered in this book) and delay-free loop elimination techniques (Zavalishin, 2018).

The filter coefficients have stayed the same in the digital domain, besides the sampling time T. The cutoff is thus:

$$\phi = \omega_c T = 2\pi F_c T$$ (7.13)

where F_c is the cutoff frequency expressed in Hz.

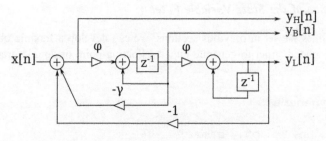

Figure 7.15: A discrete-time realization of the multimode state-variable filter.

To translate the signal flow graph into C++ code, we can look at Figure 7.15. At time step *n*, we can start from the band-pass output, looking back to the previous band-pass output and the previous high-pass output. Then we move on, evaluating the low-pass output from the current band-pass output and the previous low-pass output. Finally, we evaluate the high-pass output, taking the current input, band-pass, and low-pass outputs. The equations are rewritten in code as:

```
bp = phi*hp + bp;
lp = phi*bp + lp;
hp = x - lp - gamma*bp;
```

Good, we have derived a discrete-time formulation of the analog state-variable filter!

Before moving on to the implementation of the entire class, some details related to the coefficients should be added. As mentioned before, the discrete-time system is an approximation of the continuous-time one. This approximation leads to an error at high frequency. Specifically, the filter cutoff frequency deviates from the expected one when climbing up to the Nyquist limit. The transformation from the continuous-time to the discrete-time domain maps the frequency axis in a nonlinear fashion. This issue can be solved by inverting this map using Equation 7.14 for the computation of the cutoff coefficient:

$$\phi = 2\sin(\pi F_c/F_s). \tag{7.14}$$

Another issue related to the discrete-time implementation of the filter is stability. A high cutoff frequency and a large damping factor can lead to instability (the filter output diverges to infinity, making it unusable). Stability conditions that ensure a safe use of the filter were derived, for example, by Dattorro (2002):

$$0 \le \phi \le 1, 0 \le \gamma \le 1, \tag{7.15}$$

These conditions are slightly conservative and limit the cutoff frequency to approximately $\pi/3$ (i.e. $F_s/6$). Other stability conditions can be obtained that consider both the coefficients and allow for a higher cutoff frequency by reducing the damping (e.g. see Wise, 2006). If you are curious enough and keen to dig into the scientific literature, you will find other discretizations of the SVF or variations of the Chamberlin filter that can improve the usability range (e.g. see Wise, 2006; Zavalishin, 2018).

7.5.2 Implementation of the State-Variable Filter

Following the guidelines we set in previous sections, we can develop a flexible object that can be reused. We use a templated struct to generate either a floating-point or double-precision version. The code follows:

```
template <typename T>
struct SVF {
        T hp, bp, lp, phi, gamma;
        T fc, damp;

public:
```

```
SVF(T fc, T damp) {
        setCoeffs(fc, damp);
        reset();
}

void setCoeffs(T fc, T damp) {
                this->fc = fc;
                this->damp = damp;

                phi = clamp(2.0*sin(M_PI * fc * APP->engine->
                        getSampleTime()), 0.f, 1.f);

                gamma = clamp(2.0 * damp, 0.f, 1.f);
}

void reset() {
        hp = bp = lp = 0.f;
}

void process(T xn, T* hpf, T* bpf, T* lpf) {
        bp = *bpf = phi*hp + bp;
        lp = *lpf = phi*bp + lp;
        hp = *hpf = xn - lp - gamma*bp;
}
};
```

Following Rack guidelines, we declare it as a struct, which will make the methods and members public by default. The setCoeffs method clamps the coefficients to enforce the stability conditions.

The state-variable filter module can be designed like a wrapper for the SVF class, as it needs only to handle input and output signals and to set the filter cutoff and damping parameters. The process function follows:

```
void ASVFilter::process(const ProcessArgs &args) {

    float fc = args.sampleRate * params[PARAM_CUTOFF].getValue();

    lights[LIGHT_CLIP].value = filter->setCoeffs(fc, params[PARAM_
        DAMP].getValue());

    filter->process(inputs[MAIN_IN].getVoltageSum(), &hpf, &bpf, &lpf);

    outputs[LPF_OUT].setVoltage(lpf);
    outputs[BPF_OUT].setVoltage(bpf);
    outputs[HPF_OUT].setVoltage(hpf);
}
```

The module is monophonic, but the input handles gracefully the presence of multiple signals by summing the all in one, using the getVoltageSum method.

The reader is encouraged to read the rest of the code in the book repository.

Exercise 1

Evaluating the coefficient ϕ requires a sin(), which has a cost. We should avoid calculating the coefficients each time the process function is invoked by checking whether the input arguments have changed. Can you formulate a way to do this? This trick reduces the CPU time by at least a factor of 2 on my system. What about yours? Observe the values reported by the CPU timer with and without the trick.

Exercise 2

The cutoff of the filter is a very important parameter for the user, and the linear mapping used here is not so handy, as the most relevant frequency range is compressed in a short knob turn. Try to implement a mapping that provides unequal spacing, such as the following:

$$f_c = v^{10}, \; v \in [1.0 - 2.5] \tag{7.16}$$

where v is the knob value and the range span by the knob is approximately 10 Hz to 10 kHz.

Exercise 3

One annoyance of the formulation followed hereby is the use of a damping coefficient, which is quite unusual for musical use. Can you figure out how to replace the damping knob with a resonance or "Q" knob (i.e. inverting its value range)?

Exercise 4

The cutoff filter is an important parameter to automate by means of control signals. Add a cutoff CV input to the module, to follow LFO or EG signals.

Exercise 5

After adding the cutoff CV, you will run into an computational cost issue: if the CV signal is not constant, the CPU-preserving trick of Exercise 1 will lose its effectiveness. Indeed, if the coefficients are continuously modulated, a faster implementation of the sine function will help. We discuss this topic in Sections 10.2.2-10.2.3. Try to implement these solutions and evaluate the effect of the approximation on the CPU time and the precision of the coefficient.

7.6 Polyphonic Filter

TIP: In this section, you will learn how to handle input and output polyphonic cables.

Up to now, we have been considering ports and cables in the traditional way (i.e. as single-signal carriers). However, Rack allows you to transmit multiple signals in one cable, hence the term "polyphonic cable." This feature can be exploited in multiple ways. One could, for example, convey heterogeneous data over one cable for convenience, such as pitch, velocity, gate, and aftertouch CVs coming from Core MIDI-CV. This is totally legitimate and reduces wires on the screen. However, the dual of wire reduction is module reduction, and I think that the true power of polyphonic cables resides in this scenario. Harnessing polyphonic cables allows you to obtain instant polyphony without having to duplicate modules. This, in turn, reduces the computational burden by avoiding drawing multiple modules and performing a lot of overhead function calls.

In this section, we are going to see how simple it is to make the state-variable filter module polyphonic. The only thing we need to duplicate is the number of SVF instances. The process function needs to perform the filtering and input/output handling in a for loop. If you look at the differences between ASVFilter.cpp and APolySVFilter.cpp, you will notice some more differences related to a change of the enum names, in order to clarify which ports are polyphonic, but this is just for the sake of readability.

Please note that we are going to take advantage of the code for Exercises 2 and 4 from the previous section, and thus we are including the exponential mapping of the frequency knob and the presence of a CV for the frequency (this will be polyphonic too).

The main differences in the code are outlined in the following. The filter member is now an array of pointers to instances of SVF, instead of a single pointer. The constructor creates new instances of the SVF<float> type assigning one to each pointer, while in the monophonic module this was done directly during the declaration of filter. The polyphonic module thus looks like:

```
struct APolySVFilter: Module {
    enum ParamIds {
        PARAM_CUTOFF,
        PARAM_DAMP,
        NUM_PARAMS,
    };
    enum InputIds {
        POLY_IN,
        POLY_CUTOFF_CV,
        NUM_INPUTS,
    };
```

```
enum OutputIds {
        POLY_LPF_OUT,
        POLY_BPF_OUT,
        POLY_HPF_OUT,
        NUM_OUTPUTS,
};

enum LightsIds {
        NUM_LIGHTS,
};

SVF<float> * filter[POLYCHMAX];
float hpf, bpf, lpf;

APolySVFilter() {
        config(NUM_PARAMS, NUM_INPUTS, NUM_OUTPUTS, NUM_LIGHTS);
        configParam(PARAM_CUTOFF, 1.f, 2.5f, 2.f, "Cutoff");
        configParam(PARAM_DAMP, 0.000001f, 0.5f, 0.25f);

        for (int ch = 0; ch < POLYCHMAX; ch++)
                filter[ch] = new SVF<float>(100, 0.1);
}

void process(const ProcessArgs &args) override;
};
```

The process function now wraps in a for loop the computing of the frequency (the sum of the frequency knob and the CV input), the filtering, and writing of the outputs. The reading of the knob is done only once, before the loop. One thing to consider at this point is the different methods provided by Rack API to handle multiple channels. The getVoltage and setVoltage methods are defined using default arguments to allow getting and setting the values of each channel:

```
void setVoltage(float voltage, int channel = 0)
float getVoltage(int channel = 0)
```

In fact, up to now, we always exploited these methods without providing the channel number, thus defaulting to the first channel. As outlined in previous chapters, all cables in Rack are polyphonic by design; they just change their appearance when they carry one channel only.

The number of channels on an input port can be detected by the following:

```
int getChannels()
```

The following method sets the number of output channels, and automatically zeroes the unused channels (if any):

```
void setChannels(int channels)
```

Bearing this in mind, we can code the process function as follows:

```
void APolySVFilter::process(const ProcessArgs &args) {

    float knobFc = pow(params[PARAM_CUTOFF].getValue(), 10.f);
    float damp = params[PARAM_DAMP].getValue();

    int inChanN = std::min(POLYCHMAX, inputs[POLY_IN].getChannels());

    int ch;
    for (ch = 0; ch < inChanN; ch++) {

        float fc = knobFc +
                        std::pow(rescale(inputs[POLY_CUTOFF_CV].
getVoltage(ch), -10.f, 10.f, 0.f, 2.f), 10.f);

        filter[ch]->setCoeffs(fc, damp);

        filter[ch]->process(inputs[POLY_IN].getVoltage(ch), &hpf,
&bpf, &lpf);

        outputs[POLY_LPF_OUT].setVoltage(lpf, ch);
        outputs[POLY_BPF_OUT].setVoltage(bpf, ch);
        outputs[POLY_HPF_OUT].setVoltage(hpf, ch);
    }

    outputs[POLY_LPF_OUT].setChannels(ch + 1);
    outputs[POLY_BPF_OUT].setChannels(ch + 1);
    outputs[POLY_HPF_OUT].setChannels(ch + 1);
}
```

This is it. Please notice that to avoid unnecessary processing, we limit the number of iterations in the for loop to inChanN (i.e. the number of input channels). We could always loop over the entire 16-channel bundle; however, this would only be a waste of CPU resources. The number of active input channels is obtained from the getChannels() method. Differently, we do not care about the number of channels in the POLY_CUTOFF_CV input since this only affects the behavior of the filter cutoff.

One important tip: the transition from monophonic to polyphonic is simplified by the fact that input and output ports are polyphonic by default, but mono-compatible.

At this point of the book, this is the only polyphonic module we have developed. For a quick test of its functionalities, however, we can rely on Fundamental modules Split and Merge.

7.6.1 To Recap: Exercise with Patches

Check the online resources to see how to complete these exercises.

7.6.1.1 Simple Snare Sound Generation

A snare sound can be extremely simplified as a short burst of wideband noise (low-pass and/or band-pass filtered). To obtain such a sound, the easiest recipe requires a noise source, a quick exponential decay, a trigger for the envelope generator, and a filter. These modules are already in the ABC plugin and have been discussed extensively. You will only need to add a VCA from the Fundamental plugin.

7.6.1.2 Auto-Wah Effect

The wah-wah effect is quite popular among guitar players, but also keyboard players, especially for the rhythmic parts. The Clavinet is a very popular keyboard instrument that works well with this effect. In its essence, the wah-wah is a filter with controllable cutoff frequency. It can be of the band-pass or the low-pass type. In the latter case, the resonance must be sufficiently strong.

While guitar players are quite protective of their wah-wah pedal, many keyboard players are OK with giving away the pedalling to software automation. The so-called auto-wah effect is a variation of the effect that reacts to the input signal amplitude to change the cutoff frequency.

The auto-wah can be built from an envelope follower that can react to quick transients and an SVF with low damping. The cutoff of the filter must be controlled by a CV, as suggested by Exercise 4 in Section 7.5.

7.6.1.3 Creating a Step Filter

A step filter is a filter that abruptly changes its cutoff frequency, usually following a bpm tempo. By adding a cutoff CV input to the state-variable filter (Exercise 4 in Section 7.5) and using the random module with a slow time constant, you can obtain a step filter effect. Try it with a guitar signal or any other instrument input.

7.6.1.4 The Chemist's Lab, Aka Polyphonic Noise Filters

Although we have only developed one polyphonic module, we can take advantage of Fundamental modules Split and Merge to send multiple sources to our APolySVFilter module. Merge N noise signals from ARandom and use the resulting polyphonic cable as input to APolySVFilter. Similarly, merge N LFO signals and send them to the cutoff CV input of APolySVFilter. This will create sweeps, steam blows, and bubbles.

Note

1 Additionally, band-stop mode can be achieved by summing the low-pass and high-pass outputs.

CHAPTER 8

Crunching Numbers: Advanced DSP Modules

This section deals with some more advanced topics that may be useful to start designing Rack modules. Since the range of topics to cover is incredibly wide, three selected topics have been covered that open endless exploration:

- modal synthesis;
- virtual analog oscillators with limited aliasing; and
- waveshaping with reduced aliasing.

The latter two are surely very complex, especially in their mathematical details. For this reason, we will cover some theoretical background, but we will limit the complexity of the discussion, resorting to intuition as much as possible.

The modules in this section can get computationally expensive (see Figure 8.1). Each section discusses computational cost issues.

8.1 Modal Synthesis

 TIP: This section gives you a first look into physical modeling principles. You will also learn how to provide additional options using the context menu.

Modal synthesis (Adrien, 1991) is a sound generation technique strongly connected to physical modeling of acoustic instruments and virtual acoustics. It is based on the analysis of *modes* (i.e. the resonant frequencies of a vibrating body) and uses this knowledge to synthesize sound in an expressive and natural way. When used in physical modeling applications, the analysis of a body (a struck metal bar, a piano soundboard, etc.) is done in the time/frequency domain to characterize the number of modes, their frequency, bandwidth, and decay times. On the basis of the analysis stage, sound synthesis is performed employing resonating filter banks tuned to emulate the body resonant modes.

As with additive synthesis, the modal synthesis process can be computationally heavy, depending on the number of modes to emulate, because each mode requires one *resonator*. A resonator can be implemented by a second-order filter with tunable frequency and damping (or, inversely, Q factor).

Figure 8.1: **The modules presented in this chapter.**

Tones with hundreds of partials are better generated, employing other synthesis techniques; however, in many cases, the modal synthesis method is extremely versatile and rewarding. We shall see a simple yet powerful modal engine implementation that includes amplitude modulation to generate bell-like tones.

8.1.1 Modal Synthesis Principles

Modal synthesis is a physics-based technique that relies on the representation of an acoustic phenomenon in terms of resonant modes, and can thus be connected to Fourier series analysis/resynthesis or additive synthesis, although it departs in methods, applications, and mental framework. Generally speaking, many sound production mechanisms can be modeled as an exciter/resonator interaction, where the system is composed of an active exciter, generating the source signal, and a passive lumped resonator that shapes the source signal by imposing its own resonating modes. As an example, a struck bell can be modeled as a passive resonator (the bell), with several inharmonic resonating frequencies and an impact force, produced by the hammer, that excite these modes. The interaction between the exciter and resonator can be mutual, generating a feedback mechanism. Many types of interaction are in feedback form; however, simpler feedforward models can be sufficiently accurate. The piano string, for example, interacts with the felt hammer, producing a sequence of micro-impacts that depend on the restoration force of the hammer felt, the string displacement, and other parameters. Although this is accurately obtained with a feedback model (Bank et al., 2010), a signal-wise emulation of the impact sequence can be produced, simplifying the model and making it feedforward (Zambon et al., 2013), with computational and control advantages (Figure 8.2).

A lot of modal-based synthesis examples can be found, based on impact (Cook, 1997), friction (Avanzini et al., 2005), or blowing.

Figure 8.2: The exciter/resonator interaction is at the heart of the modal-based synthesis technique. The feedback reaction (dashed arrow) may or may not be considered in the modeling.

The advantage of modal synthesis is its flexibility and adaptability to several applications. The modeling of the exciter and resonator allows independent control of the salient timbre features. However, one of the main issues in designing a modal synthesis engine is the control of the parameters. The interaction mechanism and the exciter may have several degrees of freedom. The resonator, composed of second-order oscillators, has two degrees of freedom for each one of them. While in the emulation of physical objects a physical law can be employed to derive the parameters, in a modular synthesis setting these can be left for maximum freedom.

8.1.2 A State-Variable Filter Bank

The state-variable resonator introduced in Section 7.5 can be employed efficiently for modal synthesis. A resonant second-order filter behaves as a band-pass filter. Striking a wooden table produces a signal that has several wide bells in its spectrum. These can be obtained by a band-pass filter with low quality factor Q. Striking a string produces very sharp modes, at multiples of the fundamental frequency. These can also be obtained by a band-pass filter: if the Q factor increases, the filter bandwidth gets lower and lower, and thus its resonance gets sharp up to the point that it is almost a thin line in the frequency domain, just like a sine tone. With an increase of the Q factor, the tone produced by the filter becomes more and more sinusoidal, and its decay rate (i.e. the damping) gets lower and lower. With an infinite quality factor, the filter becomes a sinusoidal oscillator, running forever. A state-variable filter can thus be seen as a filter, a resonator, and even an oscillator. The SVF can be reused here without modifications[1] as it provides a controllable cutoff and damping band-pass mode. With zero damping, the filter can be simplified to reduce its cost and it behaves as an almost-perfect sine wave generator (see Section 10.2.4). For the current application, however, we do not want an undamped sinusoidal generator, and thus we will keep the state-variable filter previously designed as is.

8.1.3 The "Modal" Module

The module presented in this section is a simple bank of parallel second-order filters, featuring a fixed number of these. To avoid placing tens of knobs to control the cutoff and damping of each filter, each filter will be tuned with respect to a fundamental frequency, controlled by a single knob. To allow inharmonic tones to be generated, there will be an inharmonicity knob that will control the spacing of the cutoffs. The damping will be selectable (one value for all resonators), but a slope knob will also be created to increase or decrease the damping of the higher partials with respect to the lower ones. The number of active oscillators can be changed by the user from the context menu.

To increase the versatility of the model, we will also include an inexpensive amplitude modulation stage, which creates additional partials, conferring a metal-like character to the sound. The exciter of the system can be created outside the module, using other techniques that may emulate a bow friction or a sharp hit with multiple bounces. By the way, feedback interaction can be created, thanks to the architecture of VCV Rack, which allows one-sample latency between blocks.

You will also note that "hitting" the resonating system while it is still active[2] generates slightly different spectral envelopes each time, depending on the status of the filters.

The AModal module will have the following parameters:

```
enum ParamIds {
    PARAM_F0,       // fundamental frequency
    PARAM_DAMP,  // global damping
    PARAM_INHARM,    // inharmonicity
    PARAM_DAMPSLOPE,// damping slope in frequency
    PARAM_MOD_CV,    // modulation input amount
    NUM_PARAMS,
};
```

We will have two inputs, one to hit the resonators and the other one for the modulating signal:

```
enum InputIds {
    MAIN_IN,        // main input (excitation)
    MOD_IN,         // modulation input for AM
    NUM_INPUTS,
};
```

The parameters are:

- *Fundamental frequency.* The tuning knob, with logarithmic range.
- *Global damping.* All resonators, by default, will have a certain damping factor; a larger damping will provide faster decay and larger bandwidth of the resonators.
- *Damping slope.* The global damping is corrected using a damping slope, which differentiates the decay of the resonators. Specifically, the damping factor for the ith resonator will be $d + d_S \cdot i$, where d is the damping and d_s is the damping slope. For negative values of d_s, the trend is inverted: $d - d_s(N - i)$. With $d_s = 0$, all resonators will have the same damping.
- *Inharmonicity.* If set to 1, the resonators will be perfectly harmonic (i.e. at exact multiple frequencies with respect to the fundamental). In the other case, the ideal frequency of some of the oscillators will be shifted by the inharmonicity factor. We arbitrarily apply this frequency change to the even harmonics.
- *Modulation control voltage.* Sets the amount of modulation by adjusting the gain of the signal on the MOD_IN.

The architecture of the module is depicted in Figure 8.3.

The module will run a maximum of MAX_OSC (64) oscillators that are allocated during construction. The number of active oscillators, nActiveOsc, will constrain how many oscillators will run and be summed to the output. The context menu will allow the user to select between these values: 1, 16, 32, and 64.

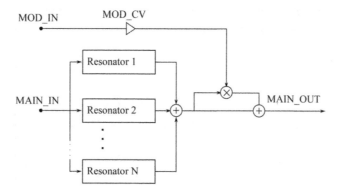

Figure 8.3: Architecture of the Modal Synthesis module, made of *N* resonators and an AM modulation stage.

Let us examine how the process function is built:

```cpp
void AModal::process(const ProcessArgs &args) {

bool changeValues = false;
float fr = pow(params[PARAM_F0].getValue(), 10.0);
        if (f0 != fr) {
        f0 = fr;
        changeValues = true;
}
if (inhrm != params[PARAM_INHARM].getValue()) {
        inhrm = params[PARAM_INHARM].getValue();
        changeValues = true;
}
if (damp != params[PARAM_DAMP].getValue()) {
        damp = params[PARAM_DAMP].getValue();
        changeValues = true;
}
if (dsl != params[PARAM_DAMPSLOPE].getValue()) {
        dsl = params[PARAM_DAMPSLOPE].getValue();
        changeValues = true;
}

float mod_cv = params[PARAM_MOD_CV].getValue();

if (changeValues) {
        for (int i = 0; i < MAX_OSC; i++) {
            float f = f0 * (float)(i+1);
            if ((i % 2) == 1)
                f *= inhrm;
            float d = damp;
            if (dsl >= 0.0)
                d += (i * dsl);
```

```
              else
                 d += ((MAX_OSC-i) * (-dsl));
              osc[i]->setCoeffs(f, d);
       }
    }

    float in = inputs[MAIN_IN].getVoltage();

    float invOut, cosOut, sinOut, cumOut = 0.0;
    for (int i = 0; i < nActiveOsc; i++) {
              osc[i]->process(in, &invOut, &cosOut, &sinOut);
              cumOut += sinOut;
    }

    if (inputs[MOD1_IN].isConnected())
              cumOut += cumOut * mod_cv * inputs[MOD1_IN].getVoltage();

    cumOut = cumOut / nActiveOsc;
    if (outputs[MAIN_OUT].isConnected())
              outputs[MAIN_OUT].setVoltage(cumOut);
}
```

The upper portion of the code is devoted to elaborating the parameters. In particular, to avoid unneeded operations, the resonator coefficients are changed only when a parameter change is detected. In that case, the changeValues Boolean variable is true and the conditional part where all the resonator coefficients are computed is executed.

At this point, the process function of each oscillator is called, getting the sinusoidal output, which is cumulated into cumOut and normalized by dividing by the number of resonators. If the modulation input is active, its signal modulates the resonator output. It is worth mentioning that if a constant value is sent to the modulation input, it acts as a gain. Finally, the signal is sent to the output.

8.1.4 Adding Context Menu Options

This module has a new feature with respect to the ones described up to this point. It introduces additional context menu items for selecting the number of oscillators. The default context menu for a module has the following items:

- Plugin;
- Preset;
- Initialize;
- Randomize;
- Disconnect Cables;
- Duplicate; and
- Delete.

We can, however, introduce additional context menu items. In this case, we will introduce a selection of oscillator numbers. Of course, this value could be assigned to a knob or a switch. However, selecting the oscillator number has more to do with CPU constraints and user mode, rather than artistical choice. One can, for example, decide to use the module as a filter (remember that the oscillator is of the SVF type), limiting the oscillators to one, or decide how many of them to use depending on their computational cost. Furthermore, muting or adding oscillators creates sudden changes in the output that may be undesired. By adding this value to the context menu, we hide it from the panel, leaving space for other functions, and we place it under the hood, to avoid distractions.

The context menu we are going to add has these additional items:

- an empty line, for spacing;
- a title for the new section, "Oscillators";
- 1;
- 16;
- 32; and
- 64.

In general, tweaking the context menu is done in two steps, roughly:

1. Create a new struct, inheriting MenuItem. This defines how the new item or items will look and interact with our module.
2. Implement the appendContextMenu method to add the required items.

We inherit MenuItem and call the new struct nActiveOscMenuItem, which contains a pointer to the modal Module, and a number equal to the oscillators this item represents. We also override its onAction method. This determines what happens when the nActiveOscMenuItem is clicked. Specifically, it calls the onOvsFactorChange from the trivial oscillator module. This changes the oversampling factor of the module:

```
struct nActiveOscMenuItem : MenuItem {
  AModal *module;
  unsigned int nOsc;
  void onAction(const event::Action &e) override{
    module->nActiveOsc = nOsc;
  }
};
```

As to the overriding of the appendContextMenu, this is part of the module widget methods, but is declared virtual, allowing us to implement its behavior. We declare it in nActiveOscMenuItem. Now that the compiler knows it will be overridden, we can implement it.

Each of these new items must be added as a child of the context menu. We add the title "Oscillators" as a MenuLabel type, a text-only item include/ui/MenuLabel.hpp. Then we add the four items, one for each oscillator number, from 1 to 64. Each item is a pointer to a new nActiveOscMenuItem and it consists of a string (the ->text field), a pointer to the trivial oscillator (necessary to take action when the item is clicked, as seen above), a number (->nOsc) that tells the application how many oscillators have been selected by the user. The item also has a right text field, which displays a checkmark "✓" or an empty string. The CHECKMARK define applies the checkmark if the input argument condition is verified.

Finally, the item is added as a child of the context menu:

```
void AModalWidget::appendContextMenu(Menu *menu) {
 AModal *module = dynamic_cast<AModal*>(this->module);

 menu->addChild(new MenuEntry);


 MenuLabel *modeLabel = new MenuLabel();
 modeLabel->text = "Oscillators";
 menu->addChild(modeLabel);

 nActiveOscMenuItem *nOsc1Item = new nActiveOscMenuItem();
 nOsc1Item->text = "1";
 nOsc1Item->module = module;
 nOsc1Item->nOsc = 1;
 nOsc1Item->rightText = CHECKMARK(module->nActiveOsc == nOsc1Item->
   nOsc);
 menu->addChild(nOsc1Item);

 nActiveOscMenuItem *nOsc16Item = new nActiveOscMenuItem();
 nOsc16Item->text = "16";
 nOsc16Item->module = module;
 nOsc16Item->nOsc = 16;
 nOsc16Item->rightText = CHECKMARK(module->nActiveOsc == nOsc16Item->
   nOsc);
 menu->addChild(nOsc16Item);

 nActiveOscMenuItem *nOsc32Item = new nActiveOscMenuItem();
 nOsc32Item->text = "32";
 nOsc32Item->module = module;
 nOsc32Item->nOsc = 32;
 nOsc32Item->rightText = CHECKMARK(module->nActiveOsc == nOsc32Item->
   nOsc);
 menu->addChild(nOsc32Item);

 nActiveOscMenuItem *nOsc64Item = new nActiveOscMenuItem();
 nOsc64Item->text = "64";
 nOsc64Item->module = module;
 nOsc64Item->nOsc = 64;
 nOsc64Item->rightText = CHECKMARK(module->nActiveOsc == nOsc64Item->
   nOsc);
 menu->addChild(nOsc64Item);

}
```

When the user clicks one of these items, onAction is called and the number of active oscillators is instantly changed. There is no need for further operations, but – just in case – we could add these to the onAction method.

8.1.5 Computational Cost

After discussing the implementation, let us assess how well it performs on a regular CPU. The computational cost may be evaluated by taking note of the number of mathematical operations conducted for each resonator, for each sample. However, a real-world measure of the execution times is sometimes more important. Execution times depend on a lot of factors, including the compiler, its flags, the processor model, the operating system, the scheduler, and so on. In the following paragraph, data is provided relative to a system with a third-generation Intel i5 processor, running a Debian-based Linux distribution.

We want to evaluate the execution times of both a compiler-optimized version and a non-optimized version. Evaluating these on a PC platform is not trivial since the system is shared by multiple processes and subject to tens of different interrupts. Fortunately, Rack provides a useful means to evaluate execution times with its CPU timer feature. The CPU timer feature provides the smoothed execution time of a single call to process and the percent ratio of the available sample time.

Table 8.1 shows the results with 1, 16, 32, and 64 oscillators, with optimization (compiler flag "-O3") and without optimization (compiler flag "-O0"). The reported values are averaged to compensate for small fluctuations. The results show that the optimization provides an increase in performance of at least three times and improves slightly with the number of oscillators because the compiler has more room to optimize.[3] The execution time does not always increase linearly with the number of oscillators. We can see, for example, that from 16 to 64 oscillators, each doubling of their number requires a 60% increase of the execution time, but the gap between 1 and 16 is 66% for the optimized case and 89% for the non-optimized case.

Please note that your platform could give very different values, depending on the CPU, the compiler, and the OS.

> **Exercise 1**
> Try computing the cost of a single oscillator in terms of sums and multiplications. Refer, for simplicity, to the flow graph of Figure 7.14c.

> **Exercise 2**
> Try to replicate the tests on your platform with optimization turned on (the default). Do not compile a non-optimized version of the plugin unless you know what you are doing.

> **Exercise 3**
> Add a method to AModal that fades out the output and then resets the resonators. Call this method when a different number of resonators is selected from the context menu.

Table 8.1: Average execution times for the modal synthesis engine, with a number of resonators varying between 1 and 64

	1	16	32	64
RT factor w/o optimization (-O0)	0.29 µs	0.55 µs	0.88 µs	1.45 µs
RT factor w/ optimization (-O3)	0.09 µs	0.15 µs	0.24 µs	0.38 µs

8.2 Virtual Analog Oscillator with Limited Aliasing

 TIP: Our first virtual analog oscillator! In this section, you will also learn how oversampling is handled in Rack. A polyphonic version will be also discussed at the end.

It is now time to deal with the scariest beast of virtual analog synthesis: aliasing! All techniques explored up to now did not pose problems of aliasing. Indeed, when designing oscillators and nonlinear components, aliasing is a serious issue that must be considered. In Chapter 2, we conducted a review of the basics about sampling and aliasing. This section discusses the implementation of a trivial oscillator design and derives two intuitive variations that reduce aliasing.

8.2.1 Trivial Oscillators

Generating classical mathematical waveforms such as the sawtooth, the square wave, or the triangle wave is pretty basic in the digital domain. The basic waveforms can be generated in the following manner:

- *Sawtooth.* Implementing a *modulo counter* (i.e. an increasing counter that resets when a threshold th_1 is hit). The modulo counter value is directly sent to the output.
- *Square wave.* Use a counter as in the sawtooth case but toggle the output each time the threshold is hit.
- *Rectangle wave with duty cycle control.* Similar to the sawtooth case but adds a second threshold $th_2 < th_1$. Outputs a high value only when the counter is larger or equal to th_2. If $th_2 = th_1/2$, the square wave is obtained as a special case (with double the frequency of the method proposed above for the square wave generation).
- *Triangle wave.* Incrementing a counter and toggling the sign of the counter increment when a threshold is surpassed.

These generation methods are generally called *trivial*, or *naïve*, as they are the simplest to implement. Figure 8.4 depicts these methods graphically.

Please note that the square wave can also be obtained as the time derivative of the triangle wave. If you require an intuitive proof of this, think about the triangle wave shape: it is composed of two segments, one with positive slope m and one with negative slope $-m$. The first-order derivative of a function, by definition, gives the slope of that function. Differentiating the trivial triangular waveform will thus give us a function that toggles between m and $-m$: a square wave!

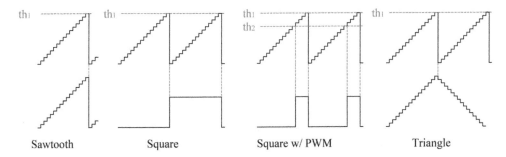

Figure 8.4: Methods for generating trivial waveforms in the digital domain. Each method shows how a counter and one or two thresholds (top line) is sufficient to generate several mathematical waveforms (bottom line).

The fundamental frequency of these waveforms can be set by either using a fixed step size and selecting the threshold accordingly, or by setting a fixed threshold and computing the required step size. For a square and a triangle wave, this method is used to determine the half-period, while for a sawtooth and a rectangle wave it determines the length of the period.

These methods are easy to implement, and it takes a few lines of code to get these waveforms generated in VCV Rack. However, as stated before, the mathematical operations described above are done in a discrete-time setting. Since we are producing a mathematical signal with infinite bandwidth in a discrete-time setting (without any additional processing), the discrete-time signal is aliased as though it was being generated in the continuous-time domain and discretized through a sampling process. Generating signals with such discontinuities thus produces aliasing that cannot be removed. Both discontinuities in the signal itself, such as the ramp reset in the sawtooth waveform, or in its first derivative, such as the slope change in the triangle wave, imply an infinite bandwidth. However, these signals have a monotonous decreasing envelope of −6 dB/oct or −12 dB/oct. It may therefore happen that the mirrored components (aliasing) are low enough not to be heard. Furthermore, psychoacoustics may alleviate the problem, because some *masking* phenomena occur that reduce our perception of aliasing. For virtual analog synthesis, we are thus not required to totally suppress aliasing. We may just reduce aliasing down to that threshold that makes it imperceptible by our auditory system. The perception of aliasing is discussed in Välimäki et al. (2010a) with experiments and data.

Let us now discuss the implementation of a trivial sawtooth. A snippet of code follows:

```
float increment = pitch / engineGetSampleRate();
moduloCounter += increment;
if ( moduloCounter > threshold) moduloCounter -= threshold;
```

In this case, the increment step is variable and the threshold is fixed. This has the advantage of obtaining signals of amplitude threshold for the whole frequency range. With a fixed step size, the amplitude would be inversely proportional to the frequency. The increment is computed in order to reach the threshold pitch times in one second (i.e. in getSampleRate() steps).

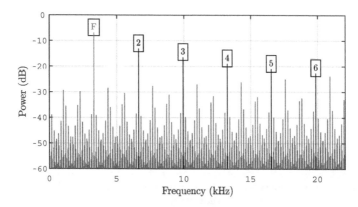

Figure 8.5: The spectrum of a 3,312 Hz sawtooth wave generated using the trivial method at 44,100 Hz. All components besides the fundamental and the harmonics up to the sixth (numbered) are undesired aliased components. The SNR is 8.13 dB, considering all aliased components as noise.

The moduloCounter is always positive. Its mean value is thus positive, introducing a DC offset, or *bias*. In the case of the sawtooth, square, and triangle waveforms, the DC offset is 0.5, and should be subtracted:

```
output = moduloCounter - 0.5;
```

As discussed previously, the trivial generation techniques run into troubles with aliasing. Figure 8.5 shows the spectrum of a 3,312 Hz sawtooth wave generated at 44,100 Hz with the trivial method in VCV Rack. Its signal-to-noise ratio (SNR) is only 8.13 dB. This means that the signal has barely 2.5 times the power of the whole aliased components! Considering that such a pitch is still inside the extension of the piano keyboard, it is unacceptable to have such a low SNR.

8.2.2 Oversampling: The Easy Solution

One easy way to reduce aliasing is oversampling. This process involves increasing the sampling rate where it matters, to reduce the effect of aliasing. It is obtained by a stage in the signal processing chain that works at a higher sampling rate (multiple integers) with respect to that of the audio engine. For an oscillator, this stage is the first in the flow graph, as shown in Figure 8.6a. The signal is generated at a high sampling rate and then converted to the audio engine sampling rate by the process of *decimation*. How does this help with aliasing? If we denote the audio engine sampling rate by F_S, as usual, and the higher sampling rate by F_H, a trivial sawtooth generated at F_H will have its spectrum mirrored around $F_H/2$, instead of $F_S/2$, as it was without oversampling. This reduces the aliasing, as shown in Figure 8.7.

Decimation consists of two steps. The bandwidth of the oversampled signal is first low-pass filtered in order to kill everything above $F_S/2$.[4] The decimation filter cuts everything over $F_S/2$. Finally, the signal can be downsampled by decimating the samples of a factor N (i.e. taking only one every N samples).

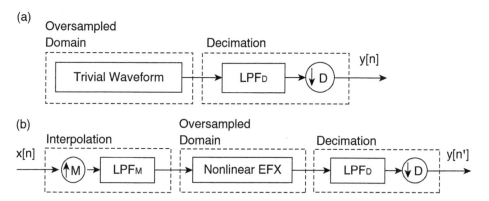

Figure 8.6: Two uses of oversampling to reduce aliasing. In (a), a trivial waveform is generated at a large sampling rate and then decimated. In (b), a nonlinear effect, generating aliasing, is run at a large sampling rate. Sampling rate conversion is done by interpolating and decimating by integer factors M and D. If $M = D$, $x[n]$ and $y[n\prime]$ are subject to the same sampling rate. If $M = D$ and no effect is applied in the oversampled domain, $x[n] = y[n\prime]$.

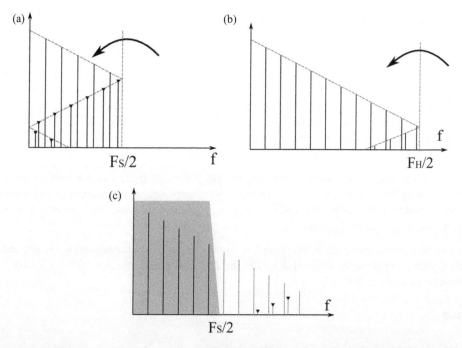

Figure 8.7: Effect of oversampling for reduced-aliasing oscillators. The effect of aliasing is shown at the sampling frequency F_S (a) and at the higher sampling frequency F_H (b). The undesired partials are highlighted by a triangle cap down to the noise floor (bottom line). If the signal is generated at F_H and decimated using the decimation low-pass filter (gray area), the sampling frequency is reverted to F_S with reduced aliasing (c).

Please note that oversampling is also used for nonlinear effects that may generate aliasing, such as distortion and waveshaping. In that case, the input signal must be first converted to the higher sampling rate by means of *interpolation* and then processed at the higher sample rate where aliasing is less of an issue, as shown in Figure 8.6b.

The use of oversampling is pretty straightforward and allows you to easily implement the trivial oscillator waveforms with reduced aliasing. However, it increases the computational requirements for two reasons:

1. The number of operations per second is increased by a factor equal to the oversampling factor.
2. The process of downsampling requires a low-pass decimation filter that needs to be as steep as possible, and is thus expensive.

8.2.3 Implementing a Trivial Oscillator with Oversampling

Rack allows easy implementation of oversampled processes by providing a decimator filter. For this reason, we will not dive into the design of such a filter, and look instead at Decimator, an object defined by Rack in Rack/include/dsp/resampler.hpp. Decimator is a templated struct that allows you to decimate with selectable decimation factor and quality factor.

```
template<int OVERSAMPLE, int QUALITY>
struct Decimator {
    . . .
};
```

The constructor of this class designs the filter impulse response and has a default input argument, 0.9, meaning that the default cutoff frequency of the filter will be $0.9 \cdot F_H/2$, but this setting can be overridden using another float input when calling the constructor. The class also has a process function, which is called to process the input vector in, of length OVERSAMPLE. The function is thus called every time we want to generate an output sample (e.g. at each call of the module process function). The higher sampling rate is $F_H = \text{OVERSAMPLE} \times F_s$.

We can now start developing a module that implements trivial generation of a waveform both at the audio engine sampling rate and at higher sampling rates. Only integer oversampling and decimation factors are considered, for simplicity. Only the sawtooth oscillator is described, leaving to you the fun of implementing other waveforms.

We want to take into consideration oversampling factors of 1× (no oversampling), 2×, 4×, and 8×. We shall define a convenience enum that makes the code more readable, where the allowed oversampling factors are defined:

```
enum {
    OVSF_1 = 1,
    OVSF_2 = 2,
    OVSF_4 = 4,
    OVSF_8 = 8,
    MAX_OVERSAMPLE = OVSF_8,
};
```

We define the struct ATrivialOsc inheriting from Module as shown:

```cpp
struct ATrivialOsc : Module {
 enum ParamIds {
      PITCH_PARAM,
      FMOD_PARAM,
      NUM_PARAMS,
 };

 enum InputIds {
      VOCT_IN,
      FMOD_IN,
      NUM_INPUTS,
 };
 enum OutputIds {
      SAW_OUT,
      NUM_OUTPUTS,
 };

 enum LightsIds {
      NUM_LIGHTS,
 };

 float saw_out[MAX_OVERSAMPLE] = {};
 float out;
 unsigned int ovsFactor = 1;
 Decimator<2,2> d2;
 Decimator<4,4> d4;
 Decimator<8,8> d8;
 int delme = 0; unsigned long cum = 0;

 ATrivialOsc() {
      config(NUM_PARAMS, NUM_INPUTS, NUM_OUTPUTS, NUM_LIGHTS);

      configParam(PITCH_PARAM, -54.0, 54.0, 0.0, "Pitch", " Hz",
        std::pow(2.f, 1.f/12.f), dsp::FREQ_C4, 0.f);
      configParam(FMOD_PARAM, 0.0, 1.0, 0.0, "Modulation");

      out = 0.0;
 }

 void process(const ProcessArgs &args) override;

 void onOvsFactorChange(unsigned int newovsf) {
      ovsFactor = newovsf;
      memset(saw_out, 0, sizeof(saw_out));
 }

};
```

The module has two inputs and two parameters. VOCT_IN is the typical V/OCT input, taking a CV to drive the pitch from, for example, a keyboard, while the FMOD_IN takes a signal for logarithmic frequency modulation of the oscillator pitch. The amount of frequency modulation is decided by a knob, FMOD_PARAM, a simple gain to be applied to the frequency modulation input. Finally, a pitch knob, PITCH_PARAM, is available, which sets the pitch when no other input is available, or offsets the pitch in the presence of non-zero inputs. Its range goes from -54 to $+54$ semitones with respect to the frequency of a C4. We shall see later how the displayBase and displayMultipler values are chosen. A float array, saw_out, is statically allocated, large enough to consider the maximum storage required (i.e. for the maximum oversampling factor). Several decimation filters are also created, d2, d4, and d8, one for each oversampling ratio. As you can see, a function is defined for setting the oversampling factor, onOvsFactorChange. The function sets the oversampling factor and cleans the saw_out array to avoid glitches during the change of the oversampling factor.

The process function follows:

```cpp
void ATrivialOsc::process(const ProcessArgs &args) {

  float pitchKnob = params[PITCH_PARAM].getValue();
  float pitchCV = 12.f * inputs[VOCT_IN].getVoltage();
  if (inputs[FMOD_IN].isConnected()) {
      pitchCV += quadraticBipolar(params[FMOD_PARAM].getValue())
        * 12.f * inputs[FMOD_IN].getVoltage();
  }
  float pitch = dsp::FREQ_C4 * std::pow(2.f, (pitchKnob + pitchCV) / 12.f);

  float incr = pitch / ((float)ovsFactor * args.sampleRate);

  if (ovsFactor > 1) {
      saw_out[0] = saw_out[ovsFactor-1] + incr;
      for (unsigned int i = 1; i < ovsFactor; i++) {
          saw_out[i] = saw_out[i-1] + incr;
          if (saw_out[i] > 1.0) saw_out[i] -= 1.0;
      }
  } else {
      saw_out[0] += incr;
      if (saw_out[0] > 1.0) saw_out[0] -= 1.0;
  }

  switch (ovsFactor) {
  case OVSF_2:
      out = d2.process(saw_out);
      break;
  case OVSF_4:
      out = d4.process(saw_out);
      break;
  case OVSF_8:
      out = d8.process(saw_out);
      break;
  case OVSF_1:
```

```
    default:
            out = saw_out[0];
            break;
    }


    if(outputs[SAW_OUT].isConnected()) {
            outputs[SAW_OUT].setVoltage(out - 0.5);
    }

}
```

The upper portion of the process function computes the pitch of the sawtooth and the increment step. The base pitch is that of a C4 (261.626 Hz). All the other contributions (i.e. frequency knob, V/OCT and FMOD inputs) add to that on a semitone scale. The base pitch is thus altered by a number of semitones equal to the sum of the pitch knob, the V/OCT input, and the FMOD input (attenuated or amplified by the frequency modulation knob).

Given a reference tone having frequency f_b, the general formula to compute the pitch of a note that is v semitones away from it is:

$$f_v = f_b \cdot 2^{v/12} \tag{8.1}$$

where v can be any real number. To clarify this, we provide a short example. Given a certain pitch (i.e. that of a C4), we obtain the pitch of a D4 by adding two semitones: $f_{D4} = f_{C4} \cdot 2^{\frac{2}{12}}$. Similarly, to get the pitch of a B3, $f_{B3} = f_{C4} \cdot 2^{\frac{-1}{12}}$.

We allow the pitch range to span nine octaves, −54 to +54 semitones with respect to the base pitch. The pitch is thus calculated as:

```
pitch = dsp::FREQ_C4 * std::pow(2.f, (pitchKnob + pitchCV) / 12.f)
```

The value displayed on the module frequency knob is calculated similarly. Its multiplier is still dsp:: FREQ_C4, while the base is $2^{1/12}$, and thus for any value of the knob the displayed result shall be $f_{C4} \cdot 2^v/12$.

For what concerns the trivial ramp generation, the increment is calculated as in the trivial case, with the exception that the increment should be divided by the oversampling factor. Indeed, the oversampled trivial waveform is computed as though we were at a higher sampling rate.

If no oversampling is applied, only one output is computed and stored in saw_out[0]. However, if oversampling is applied, the trivial algorithm is repeated in a for loop as many times as required by the oversampling factor (e.g. twice for the 2× case, four times for the 4× case, etc.). These values are stored in the saw_out array, using a part of it for 2× or 4× oversampling, or the full array in the 8× oversampling case. After computing samples at a higher sampling rate, we need to convert this signal to the engine sampling rate, using the decimator filter and yielding one sample that will go to the module output. The decimator filters all the content above $F_S/2$, so that the subsequent step (i.e. the sample rate reduction) will not add aliasing. The sample rate reduction is done by simply picking one sample over N, where N is the oversampling factor. Both the filtering process and the sampling rate reduction are conducted using one of the three Decimator filters, outputting one sample at F_S from a vector of samples at F_H. Finally, since the sawtooth wave has a DC offset, we shift the wave down by subtracting the offset.

To conclude the implementation to the module, we need to add the oversampling factor options to the context menu. We follow the same path used in Section 8.1.4 to add a separate entry for each sampling rate the module supports and define a subclass of MenuItem, called OscOversamplingMenuItem. There is one tiny difference with the Modal module. In this case, we call a method implemented in the Module, instead of directly changing a value. This method is onOvsFactorChange, and it takes care of resetting the saw_out buffer to avoid glitches.

```cpp
struct OscOversamplingMenuItem : MenuItem {
    ATrivialOsc *module;
    unsigned int ovsf;
    void onAction(const event::Action &e) override{
        module->onOvsFactorChange(ovsf);
    }
};
```

This is it. If the implementation is clear enough, we can now move on to evaluate the execution times. The evaluation follows the same criteria and code discussed in Section 8.1.3. The execution time is measured with the output port only connected to some other module and the knob in its initial position (the frequency of a C4). We also evaluate the execution times turning the frequency knob to some other values or by connecting both inputs to another module. The outcome is quite interesting. First of all, we notice that the execution times do not increase linearly with the oversampling factor. We would expect, for example, that the 8× case would take eight times more CPU time than the 1× case. This is not the case, since there is a constant cost that even the lightest modules have. For instance, the AComparator module has an execution time of 0.07 μs, even though it does not perform any tough processing. Taking this value as the average overhead of each call to the process function, we can hypothesize that the trivial sawtooth oscillator takes 0.02 μs for each cycle. Following this reasoning, the 2× oversampling implementation should take twice this time plus 0.07 μs. The same goes for the 4× and 8× cases. We can see that our hypothesis is in good accordance with the data except for a small error (0.11 μs, 0.15 μs, and 0.23 μs versus 0.11 μs, 0.14 μs, and 0.25 μs). This means that a part of the execution time increases linearly with the oversampling factor; however, a constant term is always present.

We also notice that the frequency knob value has an impact on the performance! When it is not in its initial position (0 semitones), the execution time increase largely! This is due to compiler optimizations that avoid computing the power of 2^0. In any other case (pitchKnob or pitchCV not zero), the exponentiation is computed, increasing the computational cost. There are lots of practical cases where the actual values of constants and signals impact the performance.[5]

8.2.4 Differentiated Parabolic Waveform and Its Extensions

Oversampling, as we see, is a trivial solution to the aliasing issue, and can largely increase the computational cost of a digital oscillator algorithm. Virtual analog algorithms aiming at reducing aliasing in

Table 8.2: Average execution times for the trivial sawtooth oscillator with and without oversampling

	1×	2×	4×	8×
Only output connected, knob at C4	0.09 μs	0.11 μs	0.14 μs	0.25 μs
All ports connected OR knob not at C4	0.24 μs	0.26 μs	0.3 μs	0.4 μs

smarter ways have been proposed for more than 20 years already. There is thus a large number of algorithms that may be implemented to generate reduced-aliasing VA oscillators. For this reason, selecting one for this book was not easy. Psychoacoustic tests and comparisons have been proposed for many of them (Välimäki and Huovilainen, 2007; Nam et al., 2010), suggesting that the method known as band-limited step (BLEP) (Brandt, 2001), derived from the band-limited impulse train (BLIT) (Stilson and Smith, 1996), is among the best in terms of aliasing suppression. Explaining the principles of the BLIT method and deriving the related BLEP formulation requires some basics that this book does not cover. These methods also require experiments with parameters to find an optimal trade-off. Their code is also not very linear to read and understand, since their goal is to correct the samples surrounding the steep transitions in sawtooth and square oscillators. This also makes the computational cost not very easy to determine.

An alternative method that is easier to understand and implement is the differentiated parabolic waveform method (DPW) (Välimäki, 2005) and its extensions (Välimäki et al., 2010a). We will give an intuitive explanation to the method and later consider its advantages.

We know from high school calculus classes that the integral of the derivative of a function is the function itself plus a constant value (the integral constant):

$$\int f'(x)dt = f(x) + c \tag{8.2}$$

Now, if we examine a single period of the sawtooth ramp, we can easily tell that it is a segment of a linear function of the kind $f(x) = mx + q$, where the constant term q is chosen so that the waveform has zero DC offset and the slope coefficient m sets the ramp growth time. Let us consider a wave with $q = 0$. If we integrate this linear function and we differentiate it afterwards, we will get the original function:

$$F(x) = \int f(x) = \int mxdx = \frac{m}{2}x^2 + c \quad \text{(integration step)}$$

$$\frac{dF(x)}{dx} = F'(x) = mx \quad \text{(differentiation step)} \tag{8.3}$$

Of course, it makes no sense to integrate and differentiate a function. However, it was noticed by Välimäki that if we obtain the parabolic function x^2 from squaring the ramp, instead of integrating it, and then we differentiate the parabolic function, what we obtain is a very good approximation of the sawtooth signal with the additional benefit of reduced aliasing. The alias reduction comes from the spectral envelope tilt introduced by the two operations, as shown in Figure 8.8. The benefit of the DPW algorithm is very high, even compared to the oversampling generation method seen in Section 8.2.2. The difference between a trivial and a DPW sawtooth is shown in Figure 8.9 at two different pitches. A slight correction of the sample values is responsible for such a large spectral difference.

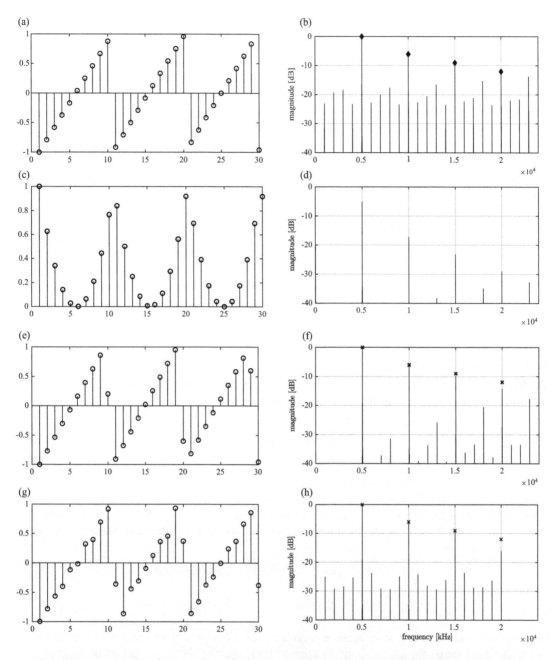

Figure 8.8: Comparison between several 5 kHz sawtooth waveforms sampled at 48 kHz. The trivial sawtooth (a), the squared sawtooth (parabolic function) (c) and its first derivative (e), and a trivial sawtooth generated with 2× oversampling (g). The respective magnitude spectra are shown on the right (b, d, f, h). The harmonics of the trivial sawtooth are highlighted in (b) with a diamond, while the rest of the partials are the effect of aliasing. The crosses in (f) and (h) show the position of the harmonics of the trivial sawtooth. A slight attenuation of the higher harmonics can be noticed by looking, for example, at the difference between the ideal amplitude (cross) of the 4th harmonic (2 kHz). While such a small difference is not noticeable, the aliasing present in (b) is largely reduced in (f). The 2× oversampling algorithm does not perform as well as the DPW algorithm and has a higher cost.

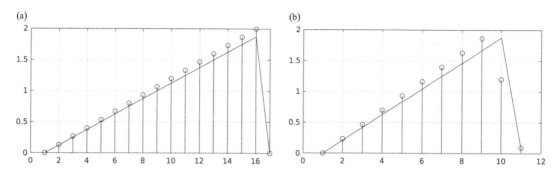

Figure 8.9: Comparison between a trivial sawtooth waveform (solid line) and the differentiated parabolic waveform (stem plot) at 3 kHz (a) and 5 kHz (b). The difference between the trivial and the alias-suppressed waveforms gets larger with increasing pitch.

The DPW algorithm is depicted in the flow graph of Figure 8.10b. Although the DPW has been shown to be not as effective as BLEP methods, it has a low code complexity, requires very low memory, and has a constant computational cost. Another advantage is the possibility of iterating the method to obtain improved aliasing reduction. It is possible, in other words, to extend the concept by finding a polynomial function of order N that provides a low aliasing sawtooth approximation when differentiated N-1 times. The general procedure to obtain such polynomials is described in Välimäki et al. (2010a) and the extension to the third order is shown in Figure 8.10c. The polynomials are reported in Table 8.3. One issue with the iterative differentiation is that the process reduces the waveform amplitude, especially at low frequency, increasing the effect of quantization by raising the signal to quantization noise ratio. This can be overcome by rescaling the waveform at each differentiation stage using a scaling factor that depends on the order of the polynomial and the period of the waveform, shown as a gain in Figure 8.10c. For DSP newbies, the amplitude reduction due to differentiation can be intuitively explained: you implement it by subtracting two contiguous values. Those values are going to be very close to each other, especially if the signal has a low frequency, and thus the output of

Table 8.3: Polynomial functions for generating a sawtooth or a triangular wave using DPW with orders from 1 to 6 and the related scaling factors to be used during differentiation

DPW Order	Polynomial Function (SAW)	Polynomial Function (TRI)	Scaling Factor
$N = 1$	x (trivial sawtooth)	$1 - 2|x|$(trivial triangle)	1
$N = 2$	x^2	$x|x| - x$	$\pi/[4\sin(\pi/P)]$
$N = 3$	$x^3 - x$	$x^3 - 3/4x$	$\pi^2/\left(6 \cdot [2\sin(\pi/P)]^2\right)$
$N = 4$	$x^4 - 2x^2$	$x^3|x| - 2x^2 + x$	$\pi^3/\left(24 \cdot [2\sin(\pi/P)]^3\right)$
$N = 5$	$x^5 - 10/3x^3 + 7/3x$	$x^5 - 5/2x^3 + 25/16x$	$\pi^4/\left(120 \cdot [2\sin(\pi/P)]^4\right)$
$N = 6$	$x^6 - 5x^4 + 7x^2$	$x^5|x| - 3x^5 + 5x^3 - 3x$	$\pi^5/\left(720 \cdot [2\sin(\pi/P)]^5\right)$

Note: $P = (F_s/f_0)$.
Table adapted from Välimäki et al. (2010a).

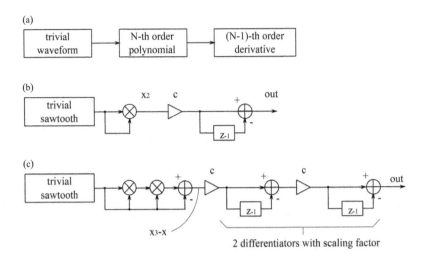

Figure 8.10: General flow graph of the DPW family of algorithms (a), the 2nd order sawtooth DPW algorithm (b), and 3rd order DPW algorithm (c).

the differentiator is generally small. Of course, the outcome of the differentiation gets large when abrupt or very quick changes happen (i.e. when contiguous samples are not similar to one another).

Several approaches have been proposed to compute the scaling factor. An accurate solution able to keep the amplitude almost constant over the oscillator frequency range is provided in Table 8.3. Some factors can be precomputed to reduce the computational cost.

The DPW technique can also be extended to the triangle wave by computing different polynomials, as shown in Table 8.3, and generating a trivial triangle wave as input. The square wave can then be obtained by differentiating the triangle wave. Table 8.3 reports the polynomials required by each DPW order for both the sawtooth and the triangular wave. Increasing the order of the DPW will increase the computational cost and reduce the aliasing further.

8.2.5 Differentiated Parabolic Waveform Implementation

Let us get to grips with the implementation of a DPW oscillator in Rack. Encapsulating the DSP code in an object is a good idea. This will also allow us to design a polyphonic version of the oscillator at no extra effort, since we just need to allocate multiple identical objects.

We first define the following two enums to improve code readability:

```
typedef enum {
    DPW_1 = 1,
    DPW_2 = 2,
    DPW_3 = 3,
    DPW_4 = 4,
    MAX_ORDER = DPW_4,
} DPWORDER;
```

```
typedef enum {
     TYPE_SAW,
     TYPE_SQU,
     TYPE_TRI,
} WAVETYPE;
```

The first four orders of the DPW are defined, and the maximum order is the fourth. We also define three types of waveform. The struct, called DPW, is templated, to allow float and double implementations, for reasons that you will discover later.

Since DPW is based on the generation of a trivial waveform, we have pitch and phase variables. The gain is the correction factor c applied during differentiation to compensate the amplitude loss due to the differentiator. The DPW order and the waveform type are stored in two variables. A buffer diffB stores the last samples, for the differentiator, and an index dbw stores the position in that buffer where the next sample will be written. This buffer is treated as circular: when the last memory element is written, the index wraps to zero, so that we start writing again from the beginning.

Finally, a variable init is used at initialization or reinitialization of the buffer to prevent glitches, as we shall see later:

```
template <typename T>
struct DPW {
  T pitch, phase;
  T gain = 1.0;
  unsigned int dpwOrder = 1;
  WAVETYPE waveType;
  T diffB[MAX_ORDER];
  unsigned int dbw = 0; // diffB write index
  int init;

  ...

}
```

The constructor sets the waveform type to the default value, clears the buffer, computes the parameters, and initializes init to the order of the DPW:

```
DPW() {
     waveType = TYPE_SAW;
     memset(diffB, 0, sizeof(T));
     paramsCompute();
     init = dpwOrder;
}
```

The paramsCompute function finds the correct scaling factor to compensate for the amplitude reduction given by the differentiator. This function is also called by setPitch, to recompute the correct gain, since the latter depends on the pitch:

```
void paramsCompute() {

        if (dpwOrder > 1)
                gain = std::pow(1.f / factorial(dpwOrder) * std::
pow(M_PI / (2.f*sin(M_PI*pitch * APP->engine->getSampleTime())),
                            dpwOrder-1.f), 1.0 / (dpwOrder-1.f));
        else
                gain=1.0;
}

void setPitch(T newPitch) {
        if (pitch != newPitch) {
                pitch = newPitch;
                paramsCompute();
        }
}
```

The generation of a the DPW sawtooth starts from the trivial sawtooth. We have a method for this where we only implement the sawtooth case – the rest is left to the reader:

```
T trivialStep(int type) {
  switch(type) {
  case TYPE_SAW:
    return 2 * phase - 1;
  default:
    return 0; // implementing other trivial waveforms is left as an
  exercise.
  }
}
```

The generation of one output sample relies on the process method, which in turn calls the trivialStep method and the dpwDiff method, used for differentiating the polynomial values. These are reported below:

```
T dpwDiff(int ord) {
  ord = clamp(ord, 0, MAX_ORDER); // avoid unexpected behavior

  T tmpA[dpwOrder];
  memset(tmpA, 0, sizeof(tmpA));
  int dbr = (dbw - 1) % ord;

  // copy last dpwOrder values into tmpA
  for (int i = 0; i < ord; i++) {
    tmpA[i] = diffB[dbr-];
    if (dbr < 0) dbr = ord - 1;
  }

  // differentiate ord-1 times
  while (ord) {
    for (int i = 0; i < ord-1; i++) {
```

```
            tmpA[i] = gain * ( tmpA[i] - tmpA[i+1] );
        }
        ord-;
    }

    return tmpA[0];
}

T process() {

    // trivial generation and update
    T triv = trivialStep(phase);
    phase += pitch * APP->engine->getSampleTime();
    if (phase >= 1.0) phase -= 1.0;

    T sqr = triv * triv;
    T poly;

    // compute the polynomial
    switch (dpwOrder) {
    case DPW_1:
    default:
        return triv;
    case DPW_2:
        poly = sqr;
        break;
    case DPW_3:
        poly = sqr * triv - triv;
        return poly;
        break;
    case DPW_4:
        poly = sqr * sqr - 2.0 * sqr;
        break;
    }

    // differentiation
    diffB[dbw++] = poly;
    if (dbw >= dpwOrder) dbw = 0;
    if (init) {
        init-;
        return poly;
    }
    return dpwDiff(dpwOrder);
}
```

The process function generates the trivial waveform and updates its phase. The polynomial is computed according to the order. Obviously, with order 1, the trivial value is just returned. The differentiation then takes place: the last value is stored into the diffB buffer and its index is advanced, with wrapping to zero. Now, if init is nonzero, we are in the very first samples, right after creation of the class or reinitialization of the DPW order. In this case, the value is just returned and the differentiation is skipped. This avoids large values, causing glitches. Indeed, if the differentiation is performed with the buffer still not filled up completely, a difference between a value and zero would occur. This would get multiplied by the scaling factor gain, possibly generating values higher than the Eurorack voltage limits. The variable init is decreased, so that when it gets to zero the section of code inside the if will not be executed anymore. At this point, the buffer is filled, and from now on the differentiation will be always performed. The buffer is as large as the highest DPW order allowed. However, only a part of it

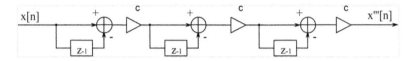

Figure 8.11: The cascade of three differentiators as required by the DPW of order 4.

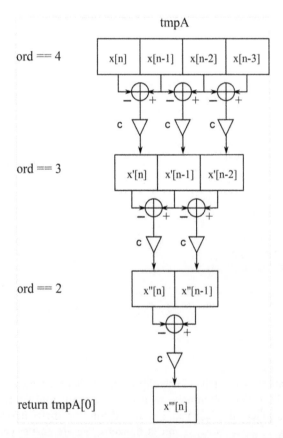

Figure 8.12: The DPW::dpwDiff method repeats several differentiation steps, as required by the algorithm. The DPW of order 4 requires differentiating three times.

will be used: the buffer is considered as large as the order. The actual size of the buffer is thus controlled by the wrapping conditions imposed to dbw. If this index gets larger than the current order, it will wrap to zero. Similarly, the number of values to wait before starting differentiating is exactly the order of the DPW (Figures 8.11 and 8.12).

As you can see, when the order changes, the code below gets executed:

```
unsigned int onDPWOrderChange(unsigned int newdpw) {
    if (newdpw > MAX_ORDER)
            newdpw = MAX_ORDER;

    dpwOrder = newdpw;
    memset(diffB, 0, sizeof(diffB));
    paramsCompute();
    init = dpwOrder;
    return newdpw;
}
```

The DPW order is set, and the buffer is erased and set to zero using memset. The gain is recomputed and init is set to the current order to allow diffB to be filled up to dpwOrder values, before starting differentiating. Finally, the new DPW order is returned. This last line is useful to return the correct value, as the check performed at the beginning of the method may alter the input argument.

This is it for the oscillator struct!

Now let's get to the module implementation. Its skeleton resembles that of the ATrivialOsc module. The module name is ADPWOsc and it will be a templated module. It has the same enums seen in ATrivialOsc and the parameters are configured similarly. The main difference is a pointer to the oscillator object, DPW. This is initialized in the constructor by dynamically allocating a DPW object. ADPWOsc also has a method to be called when changing the DPW order. This, in turn, calls the onDPWOrderChange method of the DPW object and stores the order returned by it, just in case it has been corrected.

```
template <typename T>
struct ADPWOsc : Module {
    enum ParamIds {
            PITCH_PARAM,
            FMOD_PARAM,
            NUM_PARAMS,
    };

    enum InputIds {
            VOCT_IN,
            FMOD_IN,
            NUM_INPUTS,
    };

    enum OutputIds {
            SAW_OUT,
            NUM_OUTPUTS,
    };
```

```
    enum LightsIds {
        NUM_LIGHTS,
    };


    DPW<T> *Osc;
    unsigned int dpwOrder = 1;

    ADPWOsc() {
        config(NUM_PARAMS, NUM_INPUTS, NUM_OUTPUTS, NUM_LIGHTS);
        configParam(PITCH_PARAM, -54.f, 54.f, 0.f, "Pitch", " Hz",
          std::pow(2.f, 1.f/12.f), dsp::FREQ_C4, 0.f);
        configParam(FMOD_PARAM, 0.f, 1.f, 0.f);
        Osc = new DPW<T>();
    }

    void process(const ProcessArgs &args) override;

    void onDPWOrderChange(unsigned int newdpw) {
        dpwOrder = Osc->onDPWOrderChange(newdpw); // this function
          also checks the validity of the input
    }

};
```

The process function follows. It processes the inputs and the parameters, sets the pitch, and calls the process method of the oscillator to get one output sample. Please note that to avoid unnecessary computing of coefficients, the setPitch method verifies that the pitch has not changed.

```
template <typename T> void ADPWOsc<T>::process(const ProcessArgs
&args) {

float pitchKnob = params[PITCH_PARAM].getValue();
float pitchCV = 12.f * inputs[VOCT_IN].getVoltage();
if (inputs[FMOD_IN].isConnected()) {
  pitchCV += quadraticBipolar(params[FMOD_PARAM].getValue()) * 12.f *
  inputs[FMOD_IN].getVoltage();
}
T pitch = dsp::FREQ_C4 * std::pow(2.f, (pitchKnob + pitchCV) / 12.f);

Osc->setPitch(pitch);
T out = Osc->process();

if(outputs[SAW_OUT].isConnected()) {
  outputs[SAW_OUT].setVoltage(out);
}

}
```

Similar to ATrivialOsc, where we used the context menu to choose an oversampling factor, in this case we use the context menu to pick a DPW order, allowing us to evaluate the amount of aliasing in real time. The code is very similar, but you can find it in the ABC plugin source code.

8.2.6 Computational Cost and Aliasing

We may now evaluate the performances of the DPW compared to the oversampling method. The execution times of the DPW are reported in Table 8.4 for varying order of the algorithm and for both the float and double implementations. Execution times have been evaluated according to the same method used in Section 8.1.3. As in the Atrivial module, a remarkable part of the cost is to be ascribed to the computation of the power of 2, for the pitch. Less accurate methods may be devised to spare resources. On the other hand, the cost of the DPW is much lower, compared to the oversampling method. A DPW of order 4 requires much less time than the 8× oversampling but obtains much better results. In fact, even the order 2 DPW is better than the 8× oversampling method! Figure 8.13 compares a 3,312 Hz sawtooth tone produced using these two methods. The SNR value of the 8× oversampling method is higher than that of the 2nd order DPW; however, from a perceptual standpoint, the 8× oversampling method is penalized by the presence of a floor of partials even below the fundamental frequency, which is much more tedious to the ear. All in all, the DPW method provides better results than the oversampling method, with lower computational cost.

Table 8.4: Average execution times for the DPW sawtooth oscillator

DPW order:	1	2	3	4
<float> Only output connected, knob at C4	0.07 μs	0.1 μs	0.11 μs	0.12 μs
<double> Only output connected, knob at C4	0.08 μs	0.11 μs	0.13 μs	0.15 μs
<float>All ports connected OR knob not at C4	0.18 μs	0.2 μs	0.21 μs	0.22 μs
<double> All ports connected OR knob not at C4	0.2 μs	0.22 μs	0.24 μs	0.26 μs

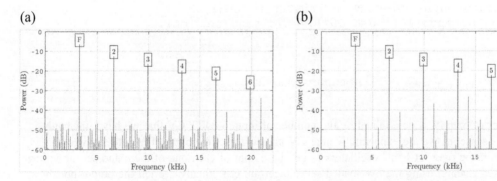

(a)
(b)

Figure 8.13: The spectrum of a 3,312 Hz sawtooth tone produced at 44,100 Hz sampling rate using 8× oversampling (a) and 2nd order DPW (b). The SNR is 22.5 dB (a) and 18 dB (b).

8.2.7 Polyphonic Oscillator Module

One of the nice things about encapsulating the DSP code of the oscillator in a class is that we can reuse it easily and give full polyphony to our module by simply changing a few lines of code!

In the monophonic module, ADPWOsc, we had a pointer to a DPW object. In ApolyDPWOsc, we can just turn it to an array of pointers:

```
DPW<T> *Osc[POLYCHMAX];
```

and we initialize each one of them in the module constructor:

```
for (int ch = 0; ch < POLYCHMAX; ch++)
    Osc[ch] = new DPW<T>();
```

The process function is slightly changed to consider multiple audio channels:

```
template <typename T> void APolyDPWOsc<T>::process (const ProcessArgs
&args) {

float pitchKnob = params[PITCH_PARAM].getValue();

int inChanN = std::min(POLYCHMAX, inputs[POLY_VOCT_IN].getChannels());

int ch;
for (ch = 0; ch < inChanN; ch++) {

    float pitchCV = 12.f * inputs[POLY_VOCT_IN].getVoltage(ch);
    if (inputs[POLY_FMOD_IN].isConnected()) {
        pitchCV += quadraticBipolar(params[FMOD_PARAM].getValue()) *
            12.f * inputs[POLY_FMOD_IN].getPolyVoltage(ch);
    }
    T pitch = dsp::FREQ_C4 * std::pow(2.f, (pitchKnob + pitchCV) / 12.f);

    Osc[ch]->setPitch(pitch);
    T out = Osc[ch]->process();

    if(outputs[POLY_SAW_OUT].isConnected()) {
        outputs[POLY_SAW_OUT].setVoltage(out, ch);
    }
}

outputs[POLY_SAW_OUT].setChannels(ch+1);
}
```

The number of outputs is determined by the number of input V/OCT CV that we have, inChanN. This makes sense, as we are expecting *N* note CV signals to set the pitch of *N* output sawtooth waves. We iterate over inChanN oscillators to get the output of each one of them and set the voltage of the related channel using the overloaded setVoltage(out, ch). Finally, we tell the output port how many channels were written using the setChannels method.

That's it! Considering that it took years to get analog engineers from the first monophonic synthesizer to the first polyphonic one, I think this two-minute job can be regarded as a considerable leap.

Exercise 1

The DPW oscillator is templated. A define at the top of ADPWOsc.cpp allows you to switch from float to double. Try compiling the code with either data type and observe with a Scope the waveform with DPW order 4 at low frequency (e.g. a C2, i.e. 65.41 Hz). What difference do you observe? What data type would you choose, in the light of the observation done and the computational cost reported above? This should also trigger a question related to design priorities when developing a synthesizer oscillator.

Exercise 2

Cheap tricks: referring to the issue observed in Exercise 1, there is one easy solution that avoids the computational burden of the double precision implementation. At low frequency, the aliasing of the trivial waveform is not noticeable. You can just implement the oscillator in single-precision float and switch to the trivial waveform below a certain frequency. Even better, you can design a transition frequency band where the trivial and the DPW oscillator output are crossfaded to ensure a smooth transition in timbre.

Exercise 3

Add the capability of generating square, rectangle, and triangle wave oscillators to the trivial and the oversampling methods.

Exercise 4

Add square and triangle waves to the DPW oscillator.

Exercise 5

The DPW oscillator makes use of iterations. This makes the code small; however, it can be shown that the cascade of several differentiators can be reduced to a feedforward filter. If implemented in such a way, the DPW has a reduced computational cost. If you wish to improve the efficiency of your code, you can implement a different function for each order of differentiation you want to support and make it as fast as possible. A graphical approach can help us find the differential equation of a cascade of differentiators and find an optimized method to compute this. The cascade of three differentiators (DPW of order 4) is shown in Figure 8.10c. The difference equation can be computed by looking at all the branches that the input signal can take. It can:

· go straight from input to output;
· pass through the first delay and then straight to output;
· go straight at the first differentiator and pass through the second delay; and
· pass through both delays.

The resulting difference equation is thus:

$$y[n] = c^2(x[n] - x[n-1] - x[n-1] + x[n-2])$$
$$= c^2(x[n] - 2x[n-1] + x[n-2]) \qquad (8.4)$$

This can be implemented in C very easily. For other DPW orders, there will be different difference equations. As a first assignment, try to get the correct difference equation for higher DPW orders. As a second assignment, implement this in code and try evaluating the computation times for the recursive and the optimized versions.

Exercise 6

Wait, we forgot to add a V/OCT input to tune the pitch of the AModal module! Now that you know how to deal with a V/OCT input, add such an input to that module.

8.3 Wavefolding

Waveshaping is one of the distinctive traits of West Coast synthesis. Waveshapers are meant to provide rich harmonic content from simple signals such as sinusoidal or triangular waveforms by applying nonlinearities. One common kind of waveshaper in West Coast synthesis is the wavefolder, or foldback. This was one of the earliest designs by Don Buchla, consisting of a device that folds back the waveform toward zero. The discontinuities that are created in the input signal at the reaching of a threshold value are determinant in the bandwidth expansion of the original signal. Unfortunately, this expansion is also a concern for virtual analog developers since the widened spectrum may fall over the Nyquist frequency and cause aliasing.

A simple wavefolder has an input/output mapping that is piecewise linear according to the relations:

$$y = f(x) = \begin{cases} \mu - (x - \mu) = 2\mu - x, & x > \mu \\ x, & -\mu \le x \le \mu \\ -\mu + (-\mu - x) = -x - 2\mu, & x < -\mu \end{cases} \qquad (8.5)$$

where μ is a threshold value. The nonlinear mapping can be seen in Figure 8.14a along with the output of such a wavefolder on a sinusoidal input in Figure 8.14b.

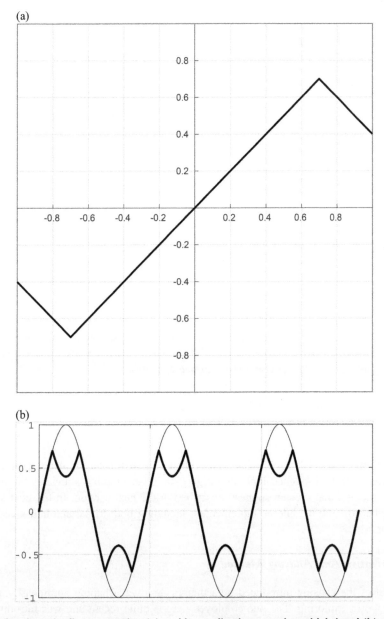

(a)

(b)

Figure 8.14: The piecewise linear mapping (a) and its application to a sinusoidal signal (b), where the thin solid line is the input sinusoidal signal and the bold line is the output of the wavefolder.

8.3.1 Trivial Foldback Algorithm

The implementation of such nonlinearity does not present particular issues in C++; it is sufficient to modify the input signal when it gets outside the linear range. Its simplest translation in code is:

```
if (x > mu)
    out = 2 * mu - x;
else if (x < -mu )
    out = - 2 * mu - x;
else
    out = x;
```

This can be refactored and shortened as:

```
if (x > mu || x < -mu)
    out = sign(x) * 2 * mu - x;
else
    out = x;
```

where we used an implementation of the sign function to contemplate both the upper and lower cases in one "if" for conciseness. Since the C++ library does not supply a sign function by default, we can implement it as a template function with inline substitution. It works on most signed data types and is branchless, to help the compiler optimize execution:

```
template <typename T> int inline sign(T val) {
    return (T(0) < val) - (val < T(0));
}
```

As we can see from Figure 8.15a, the aliasing introduced by this trivial implementation is not negligible, at least with high-pitched input and high gain. Oversampling alleviates the problem, but there is another method that we can suggest, described in the next section. In the next section, we shall also describe how the module is implemented, including both trivial and anti-aliasing implementations.

8.3.2 Antiderivative Anti-Aliasing Method

To suppress part of the aliasing introduced by a memoryless waveshaping nonlinearity, there are a few methods worth exploring. The one employed here is quite recent and was introduced in 2016 by notable names in the virtual analog research community (Parker et al., 2016). This method requires knowledge of the first antiderivative[6] $F(\cdot)$ of the nonlinear function $f(\cdot)$ in order to compute the alias-suppressed output as:

$$y_{as}[n] = \frac{F(x[n]) - F(x[n-1])}{x[n] - x[n-1]} \tag{8.6}$$

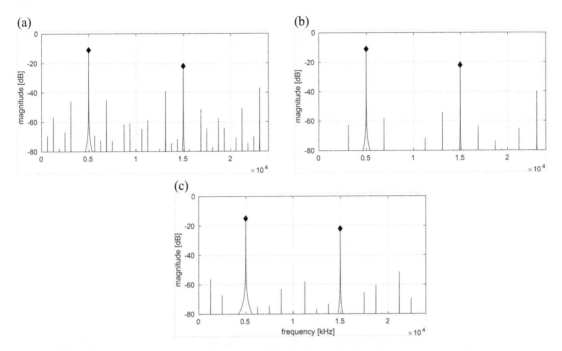

Figure 8.15: Magnitude spectra of a sine wave folded with the trivial foldback (a), the alias-reducing foldback (b), and a 2 oversampled version of the trivial foldback (c). The input signal is a 1,250 Hz sinewave with unitary gain. The foldback threshold is 0.7. The alias reduction of the oversampling method is slightly worse than the algorithm of Section 8.3.2. Please note that the foldback we described is a perfectly odd function (symmetric with respect to the origin), and thus only odd harmonics are generated.

Depending on the formulation of the function $f(\cdot)$, the computational cost of evaluating its antiderivative $F(\cdot)$ may largely vary, and thus tricks may need to be applied. Let us examine our function and proceed with integration to see how complex our $F(\cdot)$ is. Starting from Equation 8.5, we can quickly integrate the linear terms and the constant terms using basic integral relations. We must also add the integration constant, resulting in the following:

$$F(x) = \begin{cases} \frac{-x^2}{2} + 2\mu x + c_1, & x > \mu \\ \frac{x^2}{2} + c_2, & -\mu \le x \le \mu \\ \frac{-x^2}{2} - 2\mu x + c_3, & x < -\mu \end{cases} \tag{8.7}$$

To obtain a smooth function, we need to match these three curves at their endpoints. This can be done by setting this constraint and calculating the integration coefficients accordingly.

Our first constraint is that the first and the second curves need to match at the concatenation point:

$$\frac{-x^2}{2} + 2\mu x + c_1 = \frac{x^2}{2} + c_2 \quad \text{for } x = \mu \tag{8.8}$$

By plugging the value of x in the equation, we get:

$$c_1 = c_2 - \mu^2 \tag{8.9}$$

Now we only need to assign a value to c_2. If we look at Equation 8.7, we notice that the second condition is the equation of a parabola symmetrical to the vertical axis with offset c_2. We can impose that the minimum of the parabola stands in the origin by setting $c_2 = 0$. We can also notice that the third condition in Equation 8.7 is symmetrical to the first one, so it is armless to say that for our goal $c_1 = c_3$ and that $c_1 = -\mu^2$.

It is a good idea to store the $F(x[n])$ term in an auxiliary variable, so at each step the $F(x[n-1])$ need not being computed again.

8.3.3 Wavefolder Module Implementation

We can now implement the wavefolder module, allowing both the trivial and the anti-aliasing implementations to coexist, to later compare their performance with aliasing. The context menu will allow the user to evaluate both implementations, as we did with the oscillator modules.

The wavefolder will have two knobs, to adjust gain and offset. Two input ports will be associated to these two parameters, allowing CV signals to control them. Two small knobs will set the amount of these CV inputs to affect gain and offset.

By now, the definition of the module struct and its enums will look straightforward to you. The threshold is defined as a percentage of the Eurorack standard peak value.

```
#define WF_THRESHOLD (0.7f)
```

Thus, in the module constructor, we calculate the actual threshold voltage, mu, as the product of the positive peak of the Eurorack standard for oscillating signals and the above define. We also calculate its squared value to spare some computational resources later. If using the anti-aliasing method, $Fn1$ and $xn1$ will be of use. These variables store the value of $F(x[n-1])$ and $x[n-1]$, respectively.

```
struct AWavefolder : Module {
  enum ParamIds {
    PARAM_GAIN,
    PARAM_OFFSET,
    PARAM_GAIN_CV,
    PARAM_OFFSET_CV,
    NUM_PARAMS,
  };


  enum InputIds {
    MAIN_IN,
    GAIN_IN,
    OFFSET_IN,
    NUM_INPUTS,
  };
```

```
enum OutputIds {
  MAIN_OUT,
  NUM_OUTPUTS,
};

enum LightsIds {
  NUM_LIGHTS,
};

double t, tsqr;
double Fn1, xn1;
bool antialias = true;

AWavefolder() {
  config(NUM_PARAMS, NUM_INPUTS, NUM_OUTPUTS, NUM_LIGHTS);
  configParam(PARAM_GAIN_CV, 0.0, 1.0, 0.0, "Gain CV Amount");
  configParam(PARAM_OFFSET_CV, 0.0, 1.0, 0.0, "Offset CV Amount");
  configParam(PARAM_GAIN, 0.1, 3.0, 1.0, "Input Gain");
  configParam(PARAM_OFFSET, -5.0, 5.0, 0.0, "Input Offset");
  t = 5.0 * WF_THRESHOLD;
  tsqr = t*t;
  Fn1 = xn1 = 0.0;
}

void setAntialiasing(bool onOff) {
  antialias = onOff;
}

void process(const ProcessArgs &args) override;

};
```

Now let us get to the DSP part. The anti-aliasing implementation of the foldback can be implemented according to the following pseudocode:

```
if (x > mu || x < -mu)
  F = -0.5 * x*x + sign(x) * 2 * mu * x - (musqr);
else
  F = 0.5 * x*x;

out = (F - Fn1) / (x - xn1);
Fn1 = F;
xn1 = z;
```

One issue with the method we introduced is the division by the difference $x[n] - x[n-1]$, which can get very close to zero if the input signal is slowly time-varying or very low. This may introduce numerical errors or may even result in a division by zero. To avoid this ill-conditioning issue, it is advisable to adopt a different expression when the difference $x[n] - x[n-1]$ falls below a certain threshold. In this case, the output should be evaluated as:

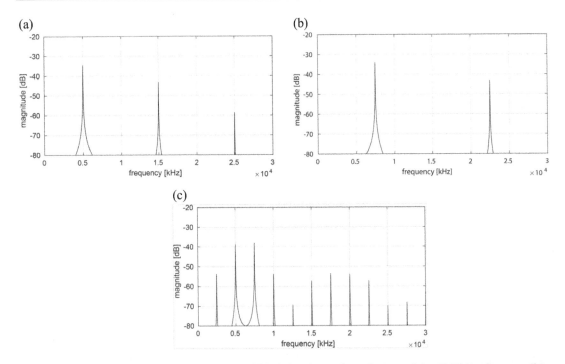

Figure 8.16: Magnitude spectra of three wavefolded signals: a 5 kHz sine tone (a), a 5.5 kHz sine tone (b), and the sum of two sine tones at 5 kHz and 5.5 kHz, both reduced by {1/2}. As can be seen, the output in (c) shows not only the harmonics of the input tones, as seen in (a) and (b), but a series of harmonics of a "ghost" 500 Hz tone (i.e. the frequency delta between the two input tones).

$$y_{as}[n] \simeq f\left(\frac{x_n + x_{n-1}}{2}\right) \tag{8.10}$$

To account for this issue, we first compute the difference and store it into a variable, dif, then decide based on its value whether to adopt Equations 8.7 or 8.10. The code follows. Please note that the output variable, out, is set at the beginning, and then replaced in case the signal is over the threshold, according to one of the techniques discussed in this chapter:

```
void AWavefolder::process(const ProcessArgs &args) {

    double offset = params[PARAM_OFFSET_CV].getValue() * inputs
[OFFSET_IN].getVoltage() / 10.0 + params[PARAM_OFFSET].getValue();
    double gain = params[PARAM_GAIN_CV].getValue() * inputs[GAIN_IN].
getVoltage() / 10.0 + params[PARAM_GAIN].getValue();
    double out, x, z;

    x = z = out = gain * inputs[MAIN_IN].getVoltage() + offset;

    if(antialias) {
        double dif = x - xn1;
```

```
    if (dif < SMALL_NUMERIC_TH && dif > -SMALL_NUMERIC_TH) {
      if (x > mu || x < -mu) {
        double avg = 0.5*(x + xn1);
        out = (sign(avg) * 2 * mu - avg);
      }
    } else {
      double F;
      if (x > mu || x < -mu)
        F = -0.5 * x*x + sign(x) * 2 * mu * x - (musqr);
      else
        F = 0.5 * x*x;
      out = (F - Fn1) / (dif);
      Fn1 = F;
    }
    xn1 = z;
  } else {
    if (x > mu || x < -mu)
      out = sign(x) * 2 * mu - x;
  }

  if (outputs[MAIN_OUT].isConnected())
    outputs[MAIN_OUT].setVoltage(out);

}
```

Exercise 1

You can try to implement a clipping nonlinearity employing the anti-aliasing method shown in this section. Integrating it is no more difficult than what we have done with the wavefolder.

Exercise 2

To alter the spectral content of the signal, this module uses a gain and an offset. The gain, however, also affects the output amplitude, which may not be desired. For instance, a large output signal could be clipped or saturated at the input of the next module. Two solutions can be devised for this issue:

1. Modify the threshold rather than the gain.
2. Multiply the output to the reciprocal of the input gain.

The latter is quick to add given the current implementation of the module. Be careful, however, to avoid dividing by zero or by very large numbers. Keeping the input gain constrained in the range 0.1–3.0, for example, works without issues.

Exercise 3

Practical wavefolding is generally realized with a cascade of several simple wavefolding stages (e.g. see the Lockhart wavefolder) (Esqueda et al., 2017). Experiment with a cascade of wavefolders with different settings, gains, or saturating nonlinearities between them. This can be done by encapsulating the wavefolder code in a separate class.

Exercise 4

As discussed above, numerical issues may arise with small values of the variable dif; this is why we proposed the use of Equation 8.10 as a solution. Let us observe what happens and when. What is the simplest periodical waveform that provides zero difference almost everywhere?

Try commenting out the "if" clause that gets to the implementation of Equation 8.10 and watch what happens with such an input signal with the anti-aliasing technique.

How can the small numerical threshold be estimated to safely avoid these issues?

8.4 To Recap: Exercise with Patches

8.4.1 A Modal Synthesis Patch

Modal synthesis modules are quite popular in the Eurorack community. These are often used to create wood, bell, or chime tones. Try to create a patch featuring:

- A modal synthesis generator.
- A sequencer driven by a clock for the V/OCT pitch input – this needs be added (see Exercise 6, Section 8.2).
- A snappy strike signal that excites the resonators, having the shape of a decaying exponential with a sharp edge at the clock transients. To make it rounder, try generating filtered noise and applying this decaying envelope to it using a VCA instead of sending it directly to the modal input.

The only module required outside the ABC plugin is a VCA to apply the decaying envelope to a filtered noise signal.

8.4.2 East Coast Classic

Now that you have all ABC modules built, try reimplementing the default VCO-VCF-VCA approach using ABC modules. Follow the default template patch that comes with VCV Rack by default and replace the Fundamental modules with ABC ones. Take note of the extra functionalities that the Fundamental modules provide and design new modules starting from the ABC ones that add these functionalities.

8.4.3 West Coast Mono Voice

The subtractive approach typical of West Coast synthesis is based on filtering a rich waveform. The East Coast approach often works the other way round: by waveshaping a dull sine or triangle wave to add overtones. In this case, the brightness of the sound is controlled by the amount of wavehsaping. The more waveshaping, the more overtones. In some cases, a filter would not even be required, as reducing the waveshaping amount reduces the richness of the output signal.

Now try to build a monophonic synthesizer employing a sine oscillator and a wavefolder instead of the traditional sawtooth/square oscillator and filter. You can start from the default Rack template patch.

8.4.4 Juntti Waveshaper

In the past, I learned a word from a Finnish friend that stayed in my mind, "Juntti." I think this is the right word to define the kind of sound we are going to design in this exercise.

Take the West Coast mono voice of the previous exercise patch and add a second sine generator. Tune it to a fifth with respect to the first oscillator (e.g. if the first is tuned to C4, tune the second to G4). Disconnect the first oscillator output. Add a Fundamental Mixer and connect both oscillators to its inputs. Now connect the Mixer output to the waveshaper and have fun mixing the levels of the two oscillators. If you look at the spectrum, you will notice that you are not only getting the higher harmonics of each sine oscillator, but also additional partials. This is shown in Figure 8.16. If f_A and f_B are the pitches of the two input sine tones, the output of the wavefolder is composed by the odd harmonics[7] of f_A ($k \cdot f_A$ for $k = 1, 3, 5, \dots$), the odd harmonics of f_B ($k \cdot f_B$ for $k = 1, 3, 5, \dots$), and the result of intermodulation: the even and odd harmonics of $f_\Delta = f_A - f_B$.

This holds true if the peak amplitude of the sum of the two sine tones is larger than the wavefolder threshold. If it does not, then the wavefolder is operating in its linear region, and thus no effect is present. This is an interesting demonstration of the difference between linear and nonlinear systems when two signals, A and B, are summed before getting processed by the system:

- When the effect is in its linear region (input always inside the positive and negative thresholds), the output corresponds exactly to the sum of A and B.
- When the effect is in its nonlinear region (input exceeding either of the thresholds), the output corresponds to the sum of the processed A and B *and* an intermodulation component that depends on both A and B.

Obviously, summing the outputs of two waveshapers (operating in their nonlinear region), receiving only one signal as input does not yield any intermodulation. In symbols: $g(A) + g(B) \neq g(A + B)$ if $g(\cdot)$ is nonlinear, or temporarily operating in a nonlinear region.

Notes

1   The only concern regards computational efficiency. By hardcoding a bank of parallel second-order filters in one class and explicitly using pragma statements to exploit processor parallelization, the computational cost can be greatly reduced. However, although often misquoted, I think Donald Knuth's statement "premature emphasis on efficiency is a big mistake" fits well in this case.

2   What in musical practice is called *ribattuto*.

3   Remember the assembly line metaphor we used in Chapter 4? It applies to the computation of batches of oscillators, not only of batches of samples.

4   Similar to the anti-aliasing filter used during the analog-to-digital conversion.

5   One example above all is the issue with denormal floating-point values (e.g. Goldberg, 1991).

6   The antiderivative is also known as the indefinite integral.

7   Please remember that the wavefolder we described in this chapter is an odd function.

CHAPTER 9

The Graphical User Interface: Creating Custom Widgets

While Chapter 5 was meant to give a quick introduction to the creation of a GUI in Rack, using ready-made widgets and allowing the reader to move quickly to the development of the first plugins, this chapter discusses the development of new widgets. We start by introducing the rendering library, NanoVG, used to draw user-defined lines and shapes and to render SVG files, then we suggest how to customize available classes to get custom knobs and text labels. At the end of the chapter, we propose a variation of the Modal Synthesis plugin that adds an interactive GUI element for sound generation.

9.1 The Rendering Library: NanoVG

The rendering library used in VCV Rack is a third-party C library called NanoVG by Mikko Mononen. NanoVG is a light and thin interface layer between OpenGL and the application core, providing 2D anti-aliased rendering only, and it has been used by several other commercial and open-source projects. The Rack GUI is designed on top of NanoVG, and each GUI component has a draw function that is called periodically to perform its fancy stuff making use of NanoVG. Among the possibilities allowed by NanoVG library, we have:

- drawing paths by nodes, arcs, beziers, and so on;
- drawing shapes (rectangles, rounded rectangles, rounds, etc.);
- striking and filling paths and shapes with solid colors or gradients;
- drawing text; and
- rendering SVG images.

All drawing operations with NanoVG are conducted referring to an object, the NVGcontext, which is passed as a pointer to each draw function and initialized by the nvgCreateGL2 function.

For instance, a draw function that implements the drawing of a rectangle is done as follows:

```
nvgBeginPath(vg);
nvgRect(vg, 100,100, 120,30);
nvgFillColor(vg, nvgRGBA(255,192,0,255));
nvgFill(vg);
```

The nvgRect defines a rectangle in the NVGcontext * vg starting at the coordinates defined by the 2nd and 3rd arguments, with width and height defined by the 4th and 5th arguments. A fill color is defined by nvgFillColor and the rectangle is filled with nvgFill. All these operations are preceded by nvgBeginPath, which must be called before each path or shape is created.

Colors are declared as NVGcolor objects, and have red, green, blue, and alpha floating-point values. However, an NVGcolor can be created using several utility functions:

- nvgRGB (or nvgRGBf for floats) taking unsigned char (or float) red, green, and blue and settings by default the alpha to the maximum (no transparency);
- nvgRGBA (nvgRGBAf) taking unsigned char (float) red, green, blue, and alpha arguments;
- NvgLerpRGBA returning a NVGcolor interpolating from two NVGcolors given as input arguments; and
- nvgHSL (nvgHSLA) taking floating point hue, saturation lightness (and alpha) as input and transforming these to a regular RGBA NVGcolor.

Moving beyond the concept of color, we have the concept of *paint* (i.e. of a varied stroke or fill). The NVGpaint object allows for gradients or patterns. Gradients can be linear, boxed, or radial, while a pattern is a repetition of an image in both the horizontal and vertical axes. Some example images are shown in Figure 9.1.

NanoVG also allows for rotations, translations, matrix transforms, scaling, and skew, making it possible to move objects and transform them.

If you have some basics of vector graphics, you can scroll through the functions of NanoVG functions. The place to start is src/nanovg.h from the NanoVG repository. However, we will examine some basic functions in this chapter. Basically, the average reader will not need to dive further with NanoVG because when inheriting a Widget or a Widget subclass he or she will only need to override the draw method and add the code he or she requires to get executed.

9.2 Custom Widgets Creation

In the following subsection, we will give an idea on the creation of widgets. We will customize the appearance of knobs and buttons, we will create text widgets, and finally we will show how to create an interactive widget that adds functionalities to the Modal Synthesis module described in

Figure 9.1: Examples of objects rendered in NanoVG: a solid color square (left) and a rounded rectangle with linear gradient paint (right).

Section 8.1.3. This widget, described in Section 9.3, has a bouncing object that responds to mouse clicks in order to "hit" the modal synthesis engine. By learning from these examples, you should be able to entirely customize your interface by changing the appearance of standard elements such as the knobs, to add informative text, or even to create touchable interfaces that react to the user.

9.2.1 Custom Knobs and Buttons

Customization of the GUI components starts with defining the appearance of knobs, buttons, and sliders, if the default ones do not match the style of the module. This is done in three steps:

1. Sketch your idea on paper and decide the width and height of the component in millimeters (mm).
2. Design an SVG image for the component with the exact size as decided at the previous step, and save it to the res/folder of your plugin.
3. Create a new struct by inheriting a basic component and define the SVG file to be used for the new component.

The reason for having the exact size decided at the beginning is that Rack does not support rescaling (it did only in versions below 0.6), and thus the size of the SVG file will also apply to the knob in Rack.

For the design of the SVG file, we take Inkscape again as the reference application. You should first open a new document and set its size. Go to File → Document Properties (or Shift+Ctrl+D) and:

- make "mm" the Default Units;
- make "mm" the Units (under Custom Size line); and
- set the Width and Height in mm as decided at step 1.

It is also suggested to use a grid, for better alignment, snapping, and positioning of objects. In the Document Properties, go to the Grid tab, create a New Rectangular Grid, and verify that the spacing for both X and Y is 1 mm and the origin is at 0, 0. Checking the "Show dots instead of lines" box is suggested for improved visibility. In Figure 9.2, we show a simple blue knob where the bounding box of the circle has been snapped to the page corners.

Exporting the drawing to SVG in the ABC/res/folder will make it available for use in our modules. Let us see how.

In Section 5.2, we reported that the RoundBlackKnob object inherits RoundKnob and adds an SVG resource, used as image. We will similarly inherit RoundKnob and add our own SVG resource to create a new class RoundBlueKnob. The code follows:

```
struct RoundBlueKnob : RoundKnob {
    RoundBlueKnob() {
    setSvg(APP->window->loadSvg(asset::system("plugins/ABC/res/blue-
knob-10.svg")));
    }
};
```

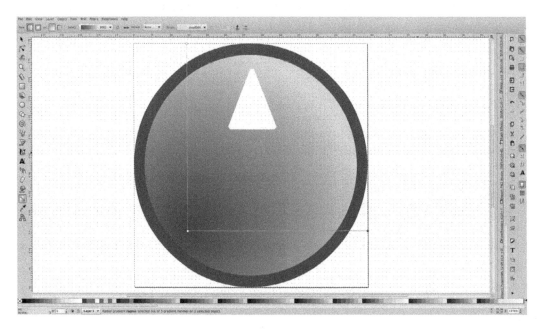

Figure 9.2: A blue knob drawn in Inkscape with size 10×10 mm.

Please note that we are using an SVG file that has size 10×10 mm. Differently, if we want to create a small trimpot, similar to the Trimpot class, we have to use a differently sized SVG, because we are not allowed to scale the SVG inside the code. We have to design a smaller knob (e.g. of size 6×6 mm).

Suppose that we also want to change the minimum and maximum angles for knob turning. We'd better inherit SvgKnob, which is a parent of RoundKnob. Indeed, the only difference between the two is that RoundKnob defines the turn angles to a fixed value, and we want to be free to set our own. We inherit from SvgKnob (defined in the app namespace) and define a new range for the knob turn:

```
struct BlueTrimpot : app::SvgKnob {
    BlueTrimpot() {
            minAngle = -0.75*M_PI;
            maxAngle = 0.75*M_PI;
            setSVG(APP->window->loadSvg(asset::system("plugins/ABC/
res/blue-knob-6.svg")));
        }
};
```

Similar considerations apply to input and output ports. A generic port graphic is defined as:

```
struct myPort : app::SvgPort {
    myPort() {
```

```
      setSVG(APP->window->loadSvg(asset::system("path/to/my/
    file.SVG"))));
        }
};
```

Please note that ports and knobs have a shadow drawn by default. The default parameters for the blur radius, the opacity, and the position can be overridden by changing the following properties in the constructor:

- *shadow->blurRadius.* The radius of the blurring (default: 0.0).
- *shadow->opacity.* The opacity of the shadow. The closer to 0.0, the more transparent (default: 0.15).
- *shadow->box.pos.* The position of the shadow. By default, it is Vec(0.0, box.size.y * 0.1) (i.e. the *x* coordinate is exactly the same as the knob, while the vertical position is slightly below by a factor dependent on the knob's *y* size). Remember that the *y* coordinate increases from top to bottom.

By changing these values, you can drop the shadow slightly to a side, blur it a bit, or change its transparency (or inversely its opaqueness). In general, making shadows consistent is a good choice for a pleasing graphical appearance. Figure 9.3 shows the default shadow setting and a shadow dropping slightly to a side with opacity 0.3 and blurRadius 2 mm.

Giving a custom look to switches follows a similar principle; however, while knobs require one image only (i.e. rotated), switches require more SVG images that are interchanged when pressing the

(a)   (b)

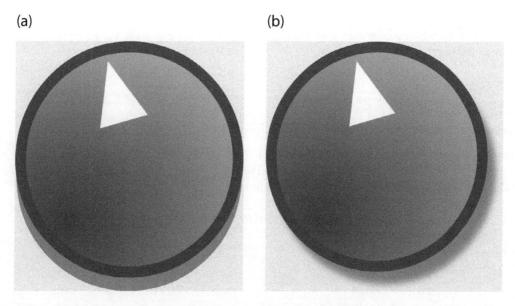

Figure 9.3: A custom knob with the default shadow settings (a) and modified shadow settings (b) (horizontal shift, opacity 0.3, and blurred shadow).

switch. These different images are called *frames*. The number of such frames should impose the number of states that the switch has. The Switch object can implement a momentary button or a maintained switch. A momentary button is active only during user pressure. A maintained switch changes its value at the user pressure. The type of switch is set by the momentary bool variable. When it is set to false (the default), more than one state is available, depending on the number of SVG frames we add to the newly defined class:

```
struct myThreeStateSwitch : app::SvgSwitch {
        myThreeStateSwitch() {
                addFrame(APP->window->loadSvg(asset::system("res/
state1.svg")));
                addFrame(APP->window->loadSvg(asset::system("res/
state2.svg")));
                addFrame(APP->window->loadSvg(asset::system("res/
state3.svg")));
        }
};

struct myPushButton : app::SvgSwitch {
        myPushButton() {
                addFrame(APP->window->loadSvg(asset::system("res/
push0.svg")));
                addFrame(APP->window->loadSvg(asset::system("res/
push1.svg")));
                momentary = true;
        }
};
```

It's as easy as that. Please note that SvgSwitch is a child of Switch that adds, for example, the pointer to the vector of the SVG frames.

9.2.2 Drawing Text

Text labels should be placed in the panel background SVG file, as already discussed. However, at times, it may be useful to add special text widgets to display informative content.

In this section, we describe a text label that reads the step of the ASequencer module and displays it. The text is placed in a rounded rectangular shape. The object inherits from TransparentWidget, a widget that does not handle mouse events. Its definition follows:

```
struct AStepDisplay: TransparentWidget {
  std::shared_ptr<Font> font;
  NVGcolor txtCol;
  ASequencer * module;
  const int fh = 20;//font height

  AStepDisplay(Vec pos);
  void setColor(unsigned char r, unsigned char g, unsigned char b,
unsigned char a);
```

```
    void draw(const DrawArgs &args) override;
    void drawBackground(const DrawArgs &args);
    void drawValue(const DrawArgs &args, const char * txt);
};
```

This class has a constant font height of 20 px, an NVGcolor struct that defines the text color, and a pointer to font type. It has a pointer to the ASequencer module for reading the step number. The setColor method sets the text color, using 8-bit values for red, green, blue, and alpha (transparency).

The draw method is the graphical equivalent of the process method for modules: Rack calls the draw method of all the instantiated widgets periodically, in order to redraw the content of each one. It is usually called with a frame rate of approximately 60 fps. When designing new widgets, we have to override the default method and add to it our custom code. The code for the draw method is:

```
void draw(const DrawArgs &args) override {
    char tbuf[2];

    if (module == NULL) return;
    snprintf(tbuf, sizeof(tbuf), "%d", module->stepNr+1);

    TransparentWidget::draw(args);
    drawBackground(args);
    drawValue(args, tbuf);
}
```

This method takes a struct containing arguments that are needed to draw. Among these, there is the pointer to the NVGcontext, found under args.vg. This is necessary, as we have seen above, for calling NanoVG methods. In our overridden draw, we first check for the existence of the module, to avoid a segmentation fault. We then read its stepNr variable, storing the current sequencer step. This is translated into text using the snprintf C library function. Finally, we call the original draw method from TransparentWidget and we supplement it with our custom drawing methods: drawBackground for the background and drawTxt for the text. The text is written after the background so that it appears over it. The functions are reported below:

```
void drawBackground(const DrawArgs &args) {
    Vec c = Vec(box.size.x/2, box.size.y);
    int whalf = box.size.x/2;
    int hfh = floor(fh/2);

    //Draw rounded rectangle
    nvgFillColor(args.vg, nvgRGBA(0xff, 0xff, 0xff, 0xF0));
    {
        nvgBeginPath(args.vg);
        nvgMoveTo(args.vg, c.x -whalf, c.y +2);
        nvgLineTo(args.vg, c.x +whalf, c.y +2);
        nvgQuadTo(args.vg, c.x +whalf +5, c.y +2+hfh, c.x +whalf, c.
y+fh+2);
        nvgLineTo(args.vg, c.x -whalf, c.y+fh+2);
```

```
          nvgQuadTo(args.vg, c.x -whalf -5, c.y +2+hfh, c.x -whalf,
    c.y +2);
          nvgClosePath(args.vg);
    }
    nvgFill(args.vg);
    nvgStrokeColor(args.vg, nvgRGBA(0x00, 0x00, 0x00, 0x0F));
    nvgStrokeWidth(args.vg, 1.f);
    nvgStroke(args.vg);
}
```

The background consists of a rounded rectangle that adds contrast to the text and highlights it. Drawing a rectangle can be done using the nvgRect command seen above. However, for this custom shape, we need to create a path point by point and fill it afterwards. To create a path, the nvgBeginPath command is issued, then a "virtual pen" is "placed on the paper" by moving its tip to a starting point with nvgMoveTo. Several nvgLineTo commands are issued to move the pen without detaching it from the paper. Finally, the nvgClosePath command tells NanoVG that the path should be closed. If you want to create curves, you have to use nvgBezierTo, nvgQuadTo, and nvgArcTo commands instead of nvgLineTo. A little knowledge on vector graphics is required to apply these functions correctly, but, as an example, we can see how to draw a rectangle with rounded sides, as in Figure 9.4. We call nvgQuadTo, in order to make the sides rounded instead of straight lines. This method requires two additional arguments, providing the coordinates of an additional point where a curved Bézier should pass through. Once the path is drawn, it can be filled (in white) and stroked (in black, with alpha 0xF).

```
    void drawValue(const DrawArgs &args, const char * txt) {
        Vec c = Vec(box.size.x/2, box.size.y);

        nvgFontSize(args.vg, fh);
        nvgFontFaceId(args.vg, font->handle);
        nvgTextLetterSpacing(args.vg, -2);
        nvgTextAlign(args.vg, NVG_ALIGN_CENTER);
```

Figure 9.4: A rectangle with rounded sides, drawn using NanoVG library. Text is superimposed by drawing it afterwards.

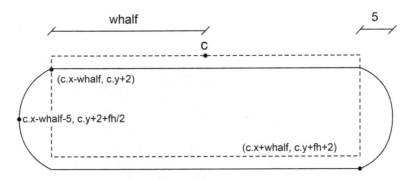

Figure 9.5: The rounded path drawn according to the code snippets provided above. The bounding box is shown in a dashed line.

```
        nvgFillColor(args.vg, nvgRGBA(txtCol.r, txtCol.g, txtCol.b,
    txtCol.a));
        nvgText(args.vg, c.x, c.y+fh, txt, NULL);
    }
```

As you can see, drawTxt calls several functions from the NanoVG library. All operations are conducted on the NanoVG context pointer vg. Font properties are set:

- the size via nvgFontSize;
- the actual font via nvgFontFaceId;
- the letter spacing nvgTextLetterSpacing is reduced to shrink a bit and fit more text; and
- the alignment is set with nvgTextAlign.

Finally, the text fill color is set and the text stored in the text string is written.

This text widget is provided as a basic didactic example to override the default draw method. You are encouraged to experiment with widgets, have a look at NanoVG, design novel concepts, and try to implement them.

9.3 An Interactive Widget

The resonator bank in AModal has an appeal for interactive applications as it can respond to any input source we may want to feed the module with. Moreover, a physical system could be displayed to enhance interactivity to stimulate user interest. We shall thus design a custom GUI that reacts to user inputs. We shall now add a module to the ABC plugin, called AModalGUI (ss Figure 9.6). It will subclass AModal and add new features. It will show a black panel where a mass can be "picked" by clicking and left falling to hit the resonator (imagine it as a bar). The GUI will thus introduce a simplified physical mechanism to simulate the mass hitting the bar and bouncing.

Let us write down the specifications for the new module:

- It will have an interactive black panel to the right.
- Interaction is done by clicking the black panel.

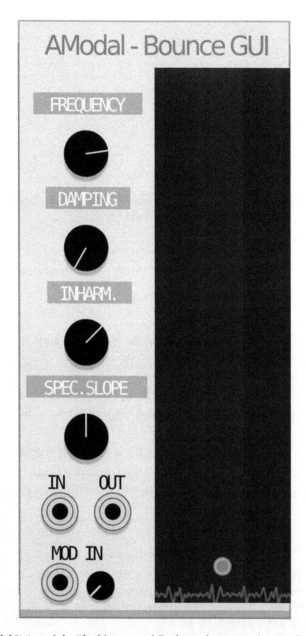

Figure 9.6: The AModalGUI module. The blue mass falls down due to gravity. By clicking on the black area, the bouncing mass is repositioned and left falling. When it hits the ground, it impacts the resonators, thus stimulating their oscillation.

- On mouse click, the mass (a blue circle) will start falling down from the set height.
- The mass will bounce as result of an anelastic collision.
- When the mass hits the ground (hit point), the velocity is transferred to the Modal module to "kick" the resonators.

- The *x* coordinate of the hit point affects the frequency slope.
- The signal resulting from the resonator output will be drawn at the hit position, resulting in a visual feedback of the impact.

The AModalGUI struct will add a new method, impact, to be called at the impact event. This is reported in Amodal.hpp. The process function will differ slightly from AModal, and thus we are rewriting it in AModalGUI.cpp. We will add a new ModuleWidget class, AModalGUIWidget, and a new widget, HammDisplay. AModalGUIWidget will contain HammDisplay, and connect it to AModalGUI, so that when an impact occurs in the GUI, the impact method from AModalGUI is called, allowing for exciting the resonators.

HammDisplay inherits from Widget, the mother of all widgets, and overrides two methods: onButton and draw. The first method is called when a click event starts (the mouse button is pressed) and will prepare variables to start the mass fall. The draw method is called periodically, and will update the display and the mass information (velocity, position, etc.), eventually calling the new impact method of AModal if the mass has touched the ground.

The declaration of HammDisplay starts as follows:

```
struct HammDisplay : Widget {
    AModalGUI *module = NULL;
    float massX;
    float massY = 0;
    float massV = 0;
    const float massA = 2.f;
    float impactY = 0;
    const float thresV = 1;
    const float massR = 0.8;
    const float massRadius = 5;
```

A pointer to the related AModal module is required, and will be initialized in AModalGUIWidget. The variables we initialized are the mass *x* and *y* coordinates, massX and massY. The velocity of the mass, massV, is zero. A constant acceleration, massA, is defined, as well as the mass radius (5 px) and a ratio, massR, later explained. A last constant value is thresV, a threshold velocity value for stopping the mass from bouncing.

The onButton event is defined as follows:

```
void onButton(const event::Button &e) override {
    e.stopPropagating();
    if (e.button == GLFW_MOUSE_BUTTON_LEFT) {
        moveMass(e.pos.x, e.pos.y);
        e.consume(NULL);
    }
}
```

This method gets the button event, e, and stop propagating to any child class, if any. If the pressed button is the left one, it calls the moveMass method, which changes the position of the mass, and consumes the event. The moveMass method just changes the mass coordinates:

```
void moveMass(float x, float y) {
      massX = x;
      massY = y;
}
```

Now, how do we ensure that the mass will fall down? The draw method will periodically update the mass position, changing its vertical position according to a constant acceleration law. We get into the draw method to see what happens at each periodical call of the draw function.

The top of draw handles the mass model:

```
void draw(NVGcontext *vg) override {

      // FALL MODEL
   if (massY <= impactY) {
           // free mass
         massV += massA;
         massY += massV;
   } else {
           // impact
         module->impact(massV, massX); // transmit velocity
         massV = -massR * massV;
         if (fabs(massV) < thresV)
                 massV = 0.f;
         else
                 massY = impactY;

   }
```

When the mass is "suspended in air" (i.e. massY is above the ground, impact), the mass is in free fall. Please note that the *y* coordinate increases from top to bottom, and thus the mass is over the hit point impactY, if its *y* value is smaller. The mass fall is simulated by simply increasing the velocity massV of a constant quantity massA, the mass acceleration, and computing the *y* coordinate by cumulating the velocity. When the mass is in touch with the resonating body (i.e. below the impactY position), we have an impact. We invoke the impact method of AModal, sending the mass velocity and *x* coordinate. We will see later how these values are used. We also invert the mass velocity vector by imposing a negative sign and reduce its absolute value to simulate a transfer of energy to the resonating body (otherwise bounces would go on indefinitely). Finally, we place the mass exactly at the position impactY to allow it be in the free mass condition at the next round. Notice that this is not a real physical model of an impact, but is intuitive and functional enough for our goals here.

To prevent infinite bounces, when the residual energy from the impact is too small (below a constant threshold), we prevent the mass from bouncing again. Specifically, if the mass velocity is below the threshold, thresV, it is stopped, by imposing a null velocity, and its position is left as is, in touch with the resonating body, preventing it from being in the free condition again.

Now, the drawing. The mass, the waveform, and the background do overlap. We thus have to draw them from the background to the foremost element, respecting their layering: the last that is drawn will cover the previous elements. The background is rendered as a dark rectangle:

```
//background
nvgFillColor(args.vg, nvgRGB(20, 30, 33));
nvgBeginPath(args.vg);
nvgRect(args.vg, 0, 0, box.size.x, box.size.y);
nvgFill(args.vg);
```

We set a dark fill color (20, 30, 33) for a rectangle of the size of the entire widget and we fill it. The variable args.vg is a pointer to NVGcontext.

The resonator is graphically represented as a waveform hinting at its oscillations: we will take the output from the module, buffer it, and draw it. The reason for buffering is that we need to draw multiple points and connect them in a path. Furthermore, the draw method is executed at a lower frequency than the process method, and thus it would be impossible to update the drawing for each output sample. AModalGUI will fill this buffer periodically and HammDisplay will read it. The waveform path is drawn, taking samples from module->audioBuffer, a float array of size SCOPE_BUFFERSIZE. The module also exposes an integer counter that tells were the last written sample was in the array. The array is treated as a circular buffer by resetting the counter to zero every time it hits the end of the array. The drawing code follows:

```
// Draw waveform
nvgStrokeColor(args.vg, nvgRGBA(0xe1, 0x02, 0x78, 0xc0));
float * buf = module->audioBuffer;
Rect b = Rect(Vec(0, 15), box.size.minus(Vec(0, 15*2)));
nvgBeginPath(args.vg);
unsigned int idx = module->idx;
for (int i = 0; i < SCOPE_BUFFERSIZE; i++) {
        float x, y;
        x = (float)i / float(SCOPE_BUFFERSIZE-1);
        y = buf[idx++];
        if (idx > SCOPE_BUFFERSIZE) idx = 0;
        Vec p;
        p.x = b.pos.x + b.size.x * x;
        p.y = impactY + y * 10.f;
        if (i == 0)
                nvgMoveTo(args.vg, p.x, p.y);
        else
                nvgLineTo(args.vg, p.x, p.y);
}
nvgLineCap(args.vg, NVG_ROUND);
nvgMiterLimit(args.vg, 2.0);
nvgStrokeWidth(args.vg, 1.5);
nvgGlobalCompositeOperation(args.vg, NVG_LIGHTER);
nvgStroke(args.vg);
```

We first set the color of the path with nvgStrokeColor. We get the audio buffer, which is collected inside AModal, and for each element of this buffer we set the y coordinate of a new point proportional to the value of the audio sample (a 10.f multiplier magnifies the value), and the x coordinate is a progressively increasing value, such that the last sample of the buffer will be drawn to the extreme right. The y value is offset by impactY to make it appear exactly where the impact point with the mass happens. Finally, the two commands used to draw lines are ngMoveTo for the first point of the shape and ngLineTo for any following element. ngLineTo will create a segment between the last point (previously set by ngLineTo or ngMoveTo) and the new given point. Remember that the path is created by a "virtual pen." To draw a line, the pen must be dragged on the paper (e.g. see ngLineTo). To move the pen without drawing, you have to lift it and move to a new position (ngMoveTo).

After the shape has been drawn, a few commands are invoked to set the line properties: the line cap, the miter limit, the stroke width (i.e. the thickness of the line), and the layer compositing. With the nvgStroke, we complete the operations and finally get the line visible.

The mass is the last element we need to draw:

```
// Draw Mass
NVGcolor massColor = nvgRGB (25, 150, 252);
nvgFillColor (args.vg, massColor);
nvgStrokeColor (args.vg, massColor);
nvgStrokeWidth (args.vg, 2);
nvgBeginPath (args.vg);
nvgCircle (args.vg, massX, massY, massRadius);
nvgFill (args.vg);
nvgStroke (args.vg);
```

The operations are similar to those seen for the other elements. We set an RGB color and set it as the fill and stroke of the object we are going to draw. We set the stroke of the object, we start a new path, and make it a circle at coordinates (massX, massY) with radius massRadius. We fill and stroke it. Done!

Now let's have a look at the changes done in AModal. The impact method is simply:

```
void impact (float v, float x) {
        hitVelocity = v;
        hitPoint = ((x / MASS_BOX_W) - 0.5) * DAMP_SLOPE_MAX;
}
```

That is, when an impact occurs, the velocity and impact x coordinate are communicated to AModalGUI. The impact x coordinate is transformed into a value that is useful to modify the damp slope (i.e. it is mapped to the range ±DAMP_SLOPE_MAX).

In the process function, the damp slope is affected by hitPoint:

```
dsl += hitPoint;
```

The hit velocity is treated as an impulse that adds with the input signal port:

```
if (hitVelocity) {
        in += hitVelocity;
        hitVelocity = 0.f;
}
```

Whenever this value is nonzero, it is added to the input and zeroed as we consumed it.

As a last note, we store values in a simple circular audio buffer:

```
audioBuffer[idx++] = cumOut;
if (idx > SCOPE_BUFFERSIZE) idx = 0;
```

The audio buffer index idx wraps when the end of the buffer memory is reached; whoever reads the buffer (the HammDisplay widget) will need to take care of wrapping too.

To conclude this section, the AModalGUIWidget has a few changes with respect to AModalWidget, in that it has a different size and it adds the HammDisplay widget as follows:

```
hDisplay = new HammDisplay();
hDisplay->module = module;
hDisplay->box.pos = Vec(15*6, 30);
float height = RACK_GRID_HEIGHT - 40;
hDisplay->box.size = Vec(MASS_BOX_W, height);
hDisplay->impactY = height - 5;
addChild(hDisplay);
```

The HammDisplay is created and its properties are set, including position, a reference to the AModalGUI module, and the impact point position.

TIP: One important tip to remember when you design custom widgets: in Rack v1, the module browser draws a preview for each module, directly calling the draw method of the module widget and its children. When doing so, it does not allocate the module, and thus the pointer to the module will be NULL. It is important to enforce a check on the module pointer: if this is NULL and we proceed to call its methods and members, we get a segmentation fault! Thus, in the draw method of your custom widgets, always perform the following check:

```
if (module == NULL) return;
```

This should be done before the module is referenced in the code. However, to make the module preview appear as close as possible to the actual implementation in the module browser, you can put the check after some static graphics is drawn. In this case, we can place the check after the background rectangle is drawn, so that, at least, the black rectangle will appear in the preview. The mass and the waveform, of course, will not show up.

Exercise 1

One concern about this implementation is related to the buffering mechanism. The draw and the process methods are concurrently working on the same variables: audioBuffer and idx. One task may preempt the other, inadvertently modifying the contents of these variables. This issue can be solved by introducing two buffers, one for writing and the other for reading. A synchronization mechanism, provided by a bool variable, tells whether the read buffer is ready for drawing. All buffers and the bool variable are instantiated in the module. The widget can access them through the pointer to the module.

The concept can be summarized as follows:

- The module writes to a buffer, writeBuffer, one sample at a time.
- When writeBuffer is complete, the module copies the data to the readBuffer and declares that it can be read, by setting a variable buffReady to true.
- The widget checks whether buffReady is true; if it is, it draws the waveform reading from readBuffer.
- When done, the widget sets back buffReady to false.
- The module does not copy writeBuffer to readBuffer if buffReady is true, because it knows that the widget has not yet finished reading the data. If the buffer is lost, this is no big issue for the graphical aspect.

Take a pen and paper, design your system, and try to prevent possible issues before starting coding! Synchronization of concurrent access to data is a delicate topic – take your time.

CHAPTER 10

Additional Topics

10.1 Debugging the Code

No matter what, you will always need to debug your code. Debugging DSP code in general is not very straightforward. Parts of your application code are event-driven – code is executed when an event happens (e.g. when a user clicks on a widget) – and some parts are called at sampling rate. The former can be debugged easily; they allow you to stop the execution at any time to inspect variable values, check the call stack, or follow the execution flow. However, with the signal processing routines, you can't always do that, as these run too fast for inspection, and sometimes getting in the way (stopping the execution) may alter the execution flow, resulting in unwanted behaviors or making bugs disappear.

If the information you need to inspect can be analyzed at key events, such as the change of a parameter, or if the event happens at a low frequency, you have options to debug the code easily. In these cases, you can stop a debugger or write the debug data to a console at key events.

Inspecting the code with the GNU Debugger (gdb) is quite easy, but let us cover some basics first.

Rack and the plugins are compiled from C++ into binaries, containing the program translated into assembler and the data (e.g. hard-coded constant values). Only the machine is able to read this stuff.

A debugger is a software application that allows the user to alter the execution of its application, for example, by stopping it into specific lines of code (using so-called *breakpoints*), inspecting the values of variables, or stopping when they are read or written (so-called *watchpoints*). Furthermore, the user is able to step the code line by line, read and alter the content of the memory, or inspect the call stack (i.e. the stack of active function calls). The debugger, however, needs extra information to help the user in his or her task, including:

• the list of symbols (variables, functions, classes), their names, and their memory address; and
• the code and its correspondence to the assembly to be executed by the machine.

Debug symbols can be generated at compile time. These are necessary to perform a meaningful debug session and are generally embedded in the compiled application binary. The vanilla Rack binary that you can download from the website does not contain debug symbols. However, the makefile that comes with the development version that you clone from Git instructs the compiler to add the debug symbols using the "-g" flag. This is added to the FLAGS variable in the compile.mk makefile include file, in the Rack root folder. Thus, the Rack binary you are now using for development has debug symbols, and similarly all the plugins you compiled have debug symbols.

10.1.1 Running a Debug Session

The debugger is an application that runs your application and controls its execution. In other words, it wraps the application (or *debugee*). To execute Rack into GDB, you can issue the following command:

```
gdb -- args ./Rack -d
```

The "--args" flag tells GDB to call the application with the arguments provided after its name, in this case the "-d" flag. The debugger will start and load Rack and its symbols. If it finds these, it will show "Reading symbols from ./Rack...done."

Now a prompt appears and you can start the debug session. If you have never used GDB, a few commands will come in handy. A non-exhaustive list is reported in Table 10.1.

You will usually set some breakpoints before you start running the application. At this point, the debugger only knows symbols from Rack. Since your plugins will be loaded dynamically by Rack after you start it, their symbols are not yet known. If you ask GDB to set a breakpoint at:

```
break ADPWOsc<double>::process
```

the debugger will tell you to "Make breakpoint pending on future shared library load?" If you are sure that you spelled the filename and the line correctly, you can answer "y". Indeed, when Rack will load the libraries, their symbols will be visible to the debugger as well, and GDB will be able to make use of them. Now you can run the debuggee:

```
run
```

Rack is running and you can interact with it as usual. Now load ADPWOsc. Rack will load it, and as soon as the process function is called for the first time GDB will stop the execution at the

Table 10.1: A list of basic GDB commands

Command	Abbreviation	Arguments	Description
run	r		Starts the application
break	b	`<filename>:` `<line>`	Sets a breakpoint at the specified line in the source code
break	b	`<class>::` `<method>`	Sets a breakpoint at the start of the method
delete	d	`<breakpoint` `number>`	Deletes a previously set breakpoint
print	p	`<variable>`	Prints the value of a variable or array
next	n		Advances by one line of code without entering into functions
step	s		Advances by one line of code possibly entering into called functions
continue	c		Resumes normal execution until next break event happens or until the end of the application life
where	/		Shows the call stack
quit	q		Stops the debugee and the debugger

beginning of it, since we set a breakpoint there. Whenever the debugger stops execution, it shows a prompt where we can type commands. The debugger will also show the line of code where the debuggee was stopped (in this case, line 51):

```
Thread 4 "Engine" hit Breakpoint 1, ADPWOsc<double>::process
  (this=0x1693600, args=..)
at src/ADPWOsc.cpp:51

51          float pitchKnob = params[PITCH_PARAM].getValue();
```

The line shown has not been executed yet, and thus the value of `pitchKnob` is not initialized. Indeed, printing it with:

```
print pitchKnob
```

will probably result in a weird value: the memory area assigned to the variable has not been cleaned or zeroed, and thus we are reading values that were written in that memory area by other processes or threads in the past. If we want to read its value, we can advance by one line with:

```
next
```

Now we can again print the value of `pitchKnob` and see the actual value, given by the knob position. Please note that the `next` and `step` commands differ in that the former jumps to the next line of code in the context, while the other enters any function call it finds in its way. For instance, doing `step` at line 51 (shown above) would enter the `getValue()` method, while next goes to line 52 of the `process` function.

Please note that you can print only variables that belong to the current context and global variables. Let us show this with an example. Set a breakpoint at the process method of DPW and issue a continue to run until you get past it:

```
break DPW<double>::process
continue
```

The execution will stop here:

```
Thread 4 "Engine" hit Breakpoint 2, DPW<double>::process
(this=0x1126540) at src/DPW.hpp:83

83        T triv = trivialStep(phase);
```

Here, you will be able to examine, for example, `triv`, but not `pitchKnob`, which now belongs to the parent function and is not in the current context. If you are not convinced, try printing the call stack:

```
where
```

The first four calls are shown here:

```
#0 DPW<double>::process (this=0x19948e0) at src/DPW.hpp:83
#1 0x00007fffec010e35 in ADPWOsc<double>::process (this=0x19da920,
    args=..) at src/ADPWOsc.cpp:59
#2 0x00000000006d9d26 in rack::engine::Engine_stepModules
    (that=0xdde060, threadId=0)
```

```
        at src/engine/Engine.cpp:251
    #3 0x00000000006da256 in rack::engine::Engine_step (that=0xdde060) at
        src/engine/Engine.cpp:334
```

As you can see, the current context is DPW<double>::process. This was called by ADPWOsc<double>::process, which in turn was called by the engine function stepModules, and so on and so forth.

Now advance through the DPW::process until you get to the return statement. Going past it will take you back to ADPWOsc::process. Once you are back there, you will be able to see pitchKnob again and all the other variables that are in the scope.

10.1.2 Debugging a Segmentation Fault

The debugger will also stop when something goes wrong (e.g. in the case of segmentation faults). A segmentation fault (in short, segfault) is a memory access violation detected by a hardware component of the processor, the memory management unit. The purpose of this mechanism is to prevent illegal access to memory areas not owned by the running application, in order to avoid any sort of damage to data and to the correct functioning of the system. Usually, when the hardware detects a segmentation fault, it notifies the operating system, which in turn sends a signal to the offending application. If this has installed a signal handler for the segfault signal, it can try to recover, otherwise the operating system usually shuts down the application abnormally. Albeit crude, this solution avoids troubles to other processes or the kernel itself.

Two common causes for segfault that you may encounter are:

- null pointer dereferencing; and
- buffer overflows.

The first case occurs when the application tries to access data through a pointer to NULL (i.e. to memory address 0x0). It is up to the developer to build a robust code that checks for this condition whenever it may occur and handles it in the proper way. In Chapter 9, we discussed how the module browser calls the draw method of the widgets to draw the previews of the modules. As we have discussed, in this particular case, the module pointer is NULL. Thus, a crash will occur due to segfault if no check on this condition is done.

The buffer overflow, on the other hand, is very common when handling buffers as we do in the audio processing field: accessing memory outside of the boundaries of an array may incur in accessing restricted memory areas.

The debugger is usually able to tell where a segfault happened and allow you to examine the call stack and the variables in the context it happened.

Let us simulate a segmentation fault condition due to a buffer overflow. You can add a nasty snippet of code such as the following to the constructor of any of the ABC modules (e.g. AComparator):

```
int crashme[1] = {0};
int i = 0;
    while (1) {
            crashme[i++] = 1;
    }
```

This code clearly causes a segmentation fault, because you will soon end trying writing "1" into a protected memory area.

Now start a debug session with GDB:

```
gdb --args ./Rack -d

run
```

If you add AComparator to the rack, as soon as the module constructor is executed, the execution will stop and you will get the following:

```
Thread 1 "Rack" received signal SIGSEGV, Segmentation fault.

0x00007fffec007b8c in AComparator::AComparator (this=0x197e810) at src/
AComparator.cpp:31

31          crashme[i++] = 1;
```

The debugger is telling you where in the code the issue is. You can print the counter i, to see how far it went before crossing a protected memory area. Printing the values of the variables in context gives valuable information.

Please note that the fault is detected only when the counter has crossed the end of the memory area reserved to the process. The offending code section may have written "1" to many other variables that belong to the process, and have thus been corrupted! In any case, you should always take care of your increment/decrement array indexes or pointers.

10.1.3 Development Mode

Do remember that Rack can be launched in *development* mode (input argument "-d"). The development mode differs from the regular *release* mode in a few regards:

- The login toolbar and the plugin manager are disabled.
- Version checking is disabled.
- Rack local directory is your current working directory and not the default (as discussed in Section 4.2).
- Logs are written to the terminal in color, not to a log file in the local Rack directory.

Launching in development mode may be particularly helpful when debugging the creation and deletion of your plugin, the interaction between it and Rack (e.g. opening a .vcv file), and so on.

10.1.4 Debugging with the Eclipse IDE

Although we have shown how to debug from the command line, Eclipse provides easy debugging in an integrated environment. This can reduce debug times.

Open Eclipse and open your project (in this case, ABC). Navigate the menu to Project → Properties. Go to Run/Debug Settings and create a new configuration. Assign a reasonable name to it (e.g. "debug Rack"). Provide the path to the Rack executable in the "C/C++ Application" field. Provide the "-d" argument in the Arguments tab. In the same tab, under "working directory," uncheck "use default" and provide the path of the folder where the Rack executable resides. Finally, go to the "Common" tab and add it to the "Debug" favorite menu, for recalling it quickly.

This should be enough to launch a debug session inside Eclipse. You can now launch the debug session by clicking on the debug button and selecting the "debug Rack" configuration.

The system should start Rack and stop at the beginning of the main, as seen in Figure 10.1. You can click on the continue button to go ahead. Rack will open up and run. Now you can add modules of yours, and add breakpoints to them. The debugger will be able to stop on these breakpoints. All the other benefits of a debugging session will be at your disposal, including watching variables or reading memory areas.

To handle the program flow, Eclipse provides the following buttons:

- *Resume.* See GDB command continue.
- *Suspend.* Suspends the application at any time, wherever it is.
- *Terminate.* Aborts execution.
- *Step into.* The equivalent of GDB command step.
- *Step over.* The equivalent of GDB command next.

Breakpoints can be added or deleted by double-clicking on the code window, on the column to the left, showing line numbers.

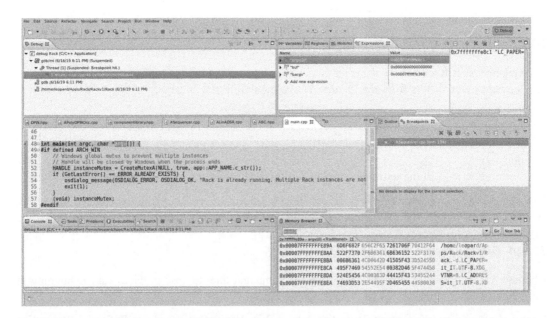

Figure 10.1: A debug session for Rack in the Eclipse IDE. On the top left, the call stack is printed, showing that the application is suspended at main() in main.cpp. The expressions window on the top right shows a list of variables or expressions and their values. The memory browser on the bottom right allows reading memory areas, such as arrays or strings. Finally, breakpoints are resumed in the right pane.

10.1.5 Other Debugging Methods

Using a debugger is very useful; however, its use is not always applicable or necessary.

We should mention that *Rack itself can be used to debug your code!* You can temporarily add ports and buttons during early development stages, similar to what hardware engineers do when they inspect a circuit, placing a probe somewhere on the PCB or soldering switches to trigger signals. Ports may allow you to extract raw signals that you can monitor with Fundamental Scope. The Scope also tells you when you get NaNs (not a number). Sometimes looking at a signal in the time domain is not sufficient. For extra information, for example, to look at aliasing, frequency of the main peak, and other spectral information, you can use a spectrum visualization module. You can also trigger events by using an additional button or input and you can use lights to monitor a Boolean status. When you set up a debug patch, you can save it and reuse it to monitor regressions or to help a collaborator spot your issue more easily.

Another possibility for debugging is *printing data to the console or to a text file.* This may be of use, for example, when a beta tester is running your plugin and he or she has no debugging skills. By printing data to a file, he or she can pass it to you for further analysis.

To print debug information with VCV Rack, you can simply use the functions provided by the logger (see include/logger.hpp):

- DEBUG (format, ...)
- INFO (format, ...)
- WARN (format, ...)
- FATAL (format, ...)

and treat these are calls to the printf C library function, as you would do in a regular C application. As an example, you may use:

```
DEBUG("value: %f", myvalue);
```

to get a printed string on the console showing the floating-point value of interest. If Rack is running in development mode, these prints will go to the console; otherwise, they are recorded into the log file.

Another option is to use `vfprintf` to print the information to a file – very useful to keep track of the values of a variable or to print the coefficients of a filter when it gets unstable (i.e. its floating-point output gets NaN).

It makes sense to print information in the constructor of your `Module` or `ModuleWidget` if you get a *segmentation fault* when Rack loads your module.

The logger functions do flush the output immediately; however, remember that `vfprintf` does not necessarily do so each time it is invoked. Usually, it does so only when the format string ends with the line termination character "`\n`," or when explicitly told to do so with the `fflush` function. In the case of the `vfprintf`, you also need to open the text file for writing (it will be created, if needed). You can do this in the constructor and close the file in the destructor. See an example application to the HelloModule module below:

```
struct HelloModule : Module {
enum ParamIds {
    NUM_PARAMS,
};
```

```
enum InputIds {
    NUM_INPUTS,
};

enum OutputIds {
    NUM_OUTPUTS,
};

enum LightsIds {
    NUM_LIGHTS,
};

FILE *dbgFile;
const char * moduleName = "HelloModule";

HelloModule() {
    config(NUM_PARAMS, NUM_INPUTS, NUM_OUTPUTS, NUM_LIGHTS);
    std::string dbgFilename = asset::user("dbgHello.log");
    dbgFile = fopen(dbgFilename.c_str(), "w");
    dbgPrint("HelloModule constructor\n");
}

~HelloModule() {
    fclose(dbgFile);
}

void process(const ProcessArgs &args) override;

void dbgPrint(const char *format, ...) {
#ifdef DEBUG
    va_list args;
    va_start(args, format);
    vfprintf(dbgFile, format, args);
    fflush(dbgFile);
    va_end(args);
#endif
    }


};
```

The resulting output, read from dbgHello.log after opening HelloModule in Rack and closing Rack, is:

```
[HelloModule]: HelloModule constructor
[HelloModule]: HelloModule destructor
```

Printing must not be abused. It is senseless to write a `printf` statement inside the `process()` function with no conditional statement because it will overload your system with data that you can't read anyway because it is printed tens of thousand times for each second. If you need to monitor a variable inside the `process()` function, you should use a counter to reduce the periodicity of execution of the `DEBUG` or `vfprintf`. An example may look as follows:

```
if (counter > args.sampleRate() * 0.5f) {
        DEBUG("%f", myVar);
        counter = 0;
} else counter++;
```

The counter variable is increased at each process invocation, and when it reaches a threshold it prints out the value of the variable `myVar`. This happens only twice for a second because we set the threshold to be half the audio sampling rate.

10.2 Optimization

Optimization is an important topic in time-critical contexts. Real-time signal processing falls into this category, and as a developer you are responsible for releasing reliable and optimized code, in order to use the least computational resources necessary for the task. Since your modules will always be part of a patch containing several other modules, the computational cost of each one has a weight on the overall performance, and all developers should avoid draining resources from other developers' modules.

Optimization, however, should not be considered prematurely. Yes, you should design your code before you start coding. You should also consider possible bottlenecks and computational issues of the implementation strategy you are adopting. You should also rely on classes and functions of the Rack library, since these are generally well thought and optimized. However, you should not waste time optimizing too early in the development process. Well-designed code requires optimization only after it has been tested and it is mature enough. Look at the CPU meter: if the module looks to be draining too many resources compared to the other ones (for a given task), then you may start optimizing.

In the optimization process, you should look for bottlenecks first. Bottlenecks are code fragments that take most of the execution time. It does not make sense to improve the speed of sections of code that do not weigh much on the overall execution time.

In general, expensive DSP tasks are those that:

- work on double-precision and complex numbers;
- iterate over large data (e.g. convolutions/FIR filters with many taps);
- nest conditional statements into iterative loops; and
- compute transcendental functions, such as exp, pow, log, sin, and so on.

Modern x86 processors have many instructions for swiftly iterating over data, working on double and complex data types, or process data in parallel. SIMD processor instructions can be exploited to improve code speed by several times. This topic, however, falls outside the scope of this book.

Other dangerous activities are those related to I/O and system calls. These operations may be blocking, and should not be called in the process method, which is time-constrained. The execution of the process function should be deterministic (i.e. the function should complete in a regular amount of time, without too much variation from time to time).

In the following subsection, we discuss methods to reduce the cost of computing transcendental function. We discuss sine and cosine approximation techniques, and table lookup for storing precomputed values of a transcendental function.

10.2.1 Table Lookup for Transcendental Functions

Lookup tables (LUTs) have a role in many musical effects and signal processing applications. Of particular interest is their use for waveshaping and wavetable synthesis. Their use can also alleviate computational issues by storing precomputed values of a transcendental function, such as exponential, trigonometric, or logarithmic functions.

Let us examine a case of moderate complexity. Suppose we have a nonlinear input/output relation that follows a Gaussian function:

$$g(x) = \frac{1}{\sigma\sqrt{2\pi}} e^{-\frac{1}{2}(x-\mu/\sigma)^2} \tag{10.1}$$

where σ and are adjustable constants, determining the horizontal offset and the width of the bell-like shape. Computing the value for each input x may be quite expensive due to the presence of several operations, including two divisions and an exponentiation.

This and many other waveshaping functions have polynomials, trigonometric series, or exponential functions that are expensive to compute on the fly. The best option is to precompute these functions at discrete input values and store them into a table. This is shown in Figure 10.2, where a set of precomputed output values (a saturating nonlinearity) are stored together with the corresponding input values. As the figure suggests, however, we do not know how to evaluate the output when the input has intermediate values not stored in the table. For instance, what will it look like?

The solution to the issue is interpolation. There are several interpolation methods. The simplest one is called *linear interpolation* and is nothing more than a weighted average between two known points. The concept is very simple. Consider the mean between two values. You compute it as $(a + b)/2$. This will give a value that is halfway between a and b (i.e. the unweighted average between the two). However, in general, we are interested in getting any intermediate value that sits

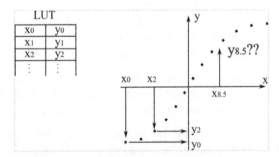

Figure 10.2: A lookup table (LUT) storing pairs of discrete input values and the corresponding output values, and its illustration. How do you compute the output for input values that have no output value precomputed in the table?

in the interval between a and b. By observing that $(a+b)/2 = 0.5a + 0.5b$, you can figure out intuitively that this is a special case for a broader range of expressions. The weighted average is thus obtained by $(1-w)a + wb$. The weight w tells where, in the interval between a and b, we should look. Let us get back to the example of Figure 10.2. If we want to evaluate the output for $y_{8.5}$, we need to get halfway between x_8 and x_9, and thus $y_{8.5} = 0.5 * x_8 + 0.5 * x_9$. If we want to get the output for input $x_{8.25}$ that is halfway between x_8 and $x_{8.5}$, we should do $y_{8.25} = 0.75x_8 + 0.25x_9$.

The approach for getting a value out of a lookup table using linear interpolation is:

$$
\begin{array}{ll}
i = \lfloor x \rfloor & (10.2a) \\
w = x - i & (10.2b) \\
\tilde{y}(x) = (1-w) \cdot y(i) + w \cdot y(i+1) & (10.2c)
\end{array}
\qquad (10.2)
$$

where i is the integer part (the brackets $\lfloor \cdot \rfloor$ denote the floor operations) and the weight w is the fractional part of i. In geometrical terms, linear interpolation essentially consists of connecting the available points with segments. Of course, this will introduce some error, as the function to be interpolated is smooth (see Figure 10.4); however, the number of operations it involves is little. Other interpolation schemes are known, which connect the points with curves instead of lines, obtaining more accurate results. There is a computational cost increase involved with these, so we will neglect them here.

Let us now discuss the practical aspects of linear interpolation. In Rack, we have a utility function, `interpolateLinear`, from include/math.hpp, which handles linear interpolation for us. This function requires the pointer to the LUT and the input value as arguments and it returns the

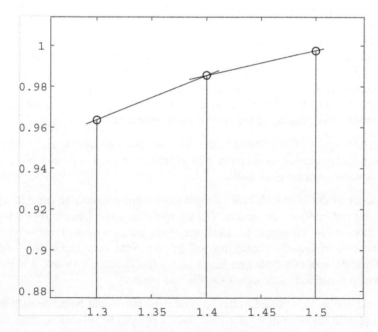

Figure 10.3: Linear interpolation for three points. Linear interpolation approximates the curve connecting the known points with segments.

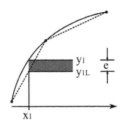

Figure 10.4: Three points of a lookup table (black dots), the curve represented by the lookup table (solid line), and the curve obtained by linear interpolation from the lookup table points. The figure shows the linear interpolation error $e = y_1 - y_{1L}$, where y_{1L} is the linearly interpolated output corresponding to input value x_1, while y_1 is the expected output for input x_1.

interpolated output value. If you therefore want to use a precomputed nonlinearity to do waveshaping, you may do something like the following:

```
float readLUT(float x) {
    #define M <length of the LUT>
    float f[M] = {<your values here>};
    x = clamp(x, -10.f, 10.f);//avoid overflow
    x = rescale(x, -10.f, 10.f, 0.f, M-1.f);//make it into an index [0, M-1]
    float y = interpolateLinear(f, x);
    return y;
}
```

What we do is define the LUT length M and its values f[0] .. f[M-1]. It is not necessary to store the values if we follow a given convention. The convention could be that M values are equally spaced in the interval $[-10; 10]$. Given this convention, we also need to ensure that the input value x never exceeds the upper and lower bounds to avoid reading outside the table boundaries. This is done using the built-in clamp function of Rack. To translate the input range into the table length (i.e. to map it into the range $[0; M - 1]$), we use the built-in function rescale. We are assuming that $x = M/2$ is the origin of the LUT. Function interpolateLinear reads the array and interpolates the value corresponding to the second input argument.

Warning: The second argument to interpolateLinear must always be lower than the size of the lookup table. Reading outside the memory area of table will yield unpredictable values and – even worse – a possible segmentation fault!

Regarding the choice of the table size, bear in mind that a trade-off must be considered between memory occupation and interpolation quality. The interpolation error depends on the number of values we store in the table: the bigger the LUT, the lower the error (noise). However, note that when defining the LUT in your C++ code, this will be part of the compiled code, increasing the size of the plugin. When the size of a table gets larger than a few thousands values, it is better to save it as a file and develop some code for loading the file into memory.

Large LUTs should not be copy-pasted in the code directly, but should be better included as external binary resources (e.g. in the form of a *.dat file). Such a file can be referenced from the source code. To do this:

1. Add the data file in the makefile:

    ```
    BINARIES += myLUT.dat
    ```

2. In the .cpp file, place the following line were you would declare the LUT:

    ```
    BINARY(myLUT_dat)
    ```

Please note that the "_" replaces the dot.

3. In the .cpp file, use the following to get the starting address, the end address, and the LUT length, respectively:

    ```
    BINARY_START(myLUT_dat)
    BINARY_END(myLUT_dat)
    BINARY_SIZE(myLUT_dat)
    ```

Please note that the first two return a (const void *) type.

For further reference, you can find these defines in include/common.hpp.

As a hint for further reads, we must also add that the table lookup technique is also used for wavetable synthesis. In this case, the table stores samples of a waveform that is read periodically to get the output waveform. Depending on the read speed, the pitch of the output waveform changes. This technique uses a playback head that can assume floating-point value, and thus requires interpolation. In Fundamental VCO (plugins/Fundamental/src/VCO.cpp), tables for the sawtooth and triangle waveforms are included. To interpolate, they use the `interpolateLinear`.

10.2.2 Sine Approximation by Table Lookup

One method to reduce the burden of evaluating a sine function is to use table lookup. The table consists of a set of precomputed values, evaluated with high precision at discrete steps of the input using a mathematical software:

$$T = \sin(x) \quad \text{for } x = \left\{0, 0.1, \ldots \frac{\pi}{2}\right\} \tag{10.3}$$

The set of points, spaced by 0.1, is shown in Figure 10.5 for the first half of a sine period. Since the remaining part of the sine period is identical apart from vertical mirroring, it is not meaningful to store values over $\pi/2$ as we can compute them from the range $0 - \pi/2$. By adding some more complexity to code even a quarter of a sine could be sufficient, since the values repeat (although also mirrored horizontally).

10.2.3 Quadratic Approximation of Sine

Another suggested method is to use an approximation formula. The Fundamental VCO (see plugins/ Fundamental/src/VCO.cpp) in Rack employs a quadratic approximation to get the sine wave output when the analog switch is turned on:

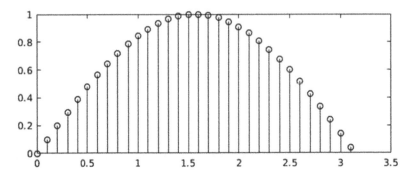

Figure 10.5: **Half a period of sine as floating-point values stored in a lookup table.**

$$\left(1 - 16\left(\phi - \tfrac{1}{4}\right)^2\right) \cdot 1.080 \leq \phi < 0.5$$
$$-\left(1 - 16\left(\phi - \tfrac{3}{4}\right)^2\right) \cdot 1.080.5 \leq \phi < 1$$

(10.4)

This sine approximation will show additional harmonics because it is not perfect, but the result is sufficient for many practical use cases.

10.2.4 The Chamberlin Resonator as a Sine Oscillator

The Chamberlin resonator, derived from the state-variable formulation of a second-order filter, is also useful for the task of generating an almost-perfect sine wave. Although we cannot use it to calculate the value of the sine and cosine trigonometric functions, we can let it spin and take its value for several tasks requiring a sine or cosine wave of arbitrary phase. As seen in Section 8.1.2, the Chamberlin resonator can provide a damped sinusoidal response when its frequency response gets very narrow. If the damping goes to zero, however, the filter never loses energy. The recursive branch with the -2ρ multiplier is now gone, reducing the computational cost of this solution, resulting in the following code:

```
hp = xn - lp;
bp = F*hp + bp;
lp = F*bp + lp;
```

For this oscillator, the output that we previously associated to the low-pass filter is now a sinusoidal output, while the output previously associated to the band-pass filter is now a cosinusoidal output.

A large number of low-cost sine oscillators have been proposed in the scientific literature, as well as quadrature sine/cosine oscillators. Some of these excel in numerical properties such as stability to coefficients change, a property that is necessary for frequency modulation, for example. Others have good precision at extremely low frequency. For the eager reader, I suggest comparing different topologies and evaluating the best ones, depending on your needs. The second-order digital waveguide oscillator is discussed in Smith and Cook (1992). The direct-form oscillator and variations of the coupled-form oscillator are shown in Dattorro (2002). Lastly, a white paper by Martin Vicanek supplies a brief comparison of several sine oscillator algorithms (Vicanek, 2015).

10.2.5 Reducing the GUI Processing Burden

The two major sources of CPU load are the modules' process function (not strange) and the GUI drawing. For what concerns the latter, something can be done to reduce its burden if your machine is not very powerful. First of all, Rack tries to shift all the graphic load to the GPU if present. You may first need to ensure that if your system has a graphic card installed capable of OpenGL2, it is working and the drivers are updated. If there are no dedicated resources for graphics processing, than all the workload is left to the CPU.

In extreme cases, you can reduce the video refresh rate, by reducing the maximum frame rate in the settings.json file at the following line:

```
"frameRateLimit": 70.0
```

A value of 25 may be a compromise between usability and performance. The graphics processing on a patch of about 15 modules on an old 64-bit processor with no GPU can be heavy. In a test on an i5 core of third generation, I get the CPU from 70% to 50% by reducing the frame rate from 70 to 25 fps.

10.3 Developing Expander Modules

Expander modules have been briefly discussed in Chapter 3. Their use is somewhat limited as they only improve the functionalities of a "parent" module. Let us discuss how these are implemented and then we will move on to the development of a simple expander that splits the polyphonic output of the APolyDPWOsc module into monophonic outputs.

The `Module` struct has pointers to modules that occur to the left and to the right, and in contact with it, so no gaps are allowed. These pointers are `leftExpander` and `rightExpander`, respectively, and are NULL if no module is to the side or if there are gaps between the two. These are pointers to the `Expander` struct, storing a pointer to a module and its ID. Furthermore, `Expander` features a messaging system at the core of the expander concept: two modules can exchange data of any type at sampling rate, in a producer/consumer fashion, by writing and reading on two independent data structures. Data structures can be of any kind, according to the developer necessities, and they are pointed by two void pointers: `producerMessage` and `consumerMessage`. The former is intended for writing and the latter for reading. Thus, once the producer module has written to `producerMessage`, it sets the Boolean variable `messageFlipRequested` to true, in order to tell the system that this can be consumed by the consumer. Let us go through an example. The mechanism is conceived such that:

- The constructor of A is executed:
  - Module A instantiates two identical data structures s_1, s_2, for now empty.

- At time step 0, module A::process is called:
  - Module A makes `producerMessage` point to s_1 and `consumerMessage` point to s_2.
  - Module A writes data to `producerMessage`.
  - Module A sets `messageFlipRequested` to true.

- At time step 0, module B::process is called:
  - Module B reads data from the `consumerMessage` pointer of module A; this is empty at the moment.

- At the beginning of a new time step, Rack examines `messageFlipRequested` and swaps the two pointers if it is true; `producerMessage` now points to s_2 and `consumerMessage` points to s_1.
- At time step 1, module A::process is called:
  - Module A writes data to producerMessage.
  - Module A sets messageFlipRequested to true.
- At time step 1, module B::process is called:
  - Module B reads data from the consumerMessage pointer of module A; this corresponds to s_1, which contains data from time step 0.
- And so on.

In this example, we considered – for simplicity – a case with module A writing data to module B. However, the mechanism allows both modules to read and write. The ping-pong mechanism implied by the two swapping pointers allows one to read on a data structure referenced by `consumerMessage`, while the other is writing to the second data structure, referenced by `producerMessage`. The two are then swapped, and thus the modules can always refer to the same pointer transparently. One obvious consequence of this mechanism is a one-sample delay between the writing and the reading of messages. Whether this is acceptable for your application is up to you. Consider that a one-sample delay is implied by connecting modules through regular cables anyway.

Now we shall consider, as an example, the implementation of an expander module that exposes all 16 individual outputs of the APolyDPWOsc. These are the specifications:

- The expander should host 16 output ports.
- The data structure that is passed to the expander is defined as an array of 16 float values.
- No data is sent back from the expander.

The data structure is defined as:

```
struct xpander16f {
    float outs[16];
};
```

Its definition should be available to both APolyDPWOsc and its expander, APolyXpander.

Let us consider, first, the changes that we have to apply to APolyDPWOsc.

Two xpander16f structures are allocated statically, for simplicity,[1] in APolyDPWOsc:

```
xpander16f xpMsg[2];
```

In the constructor, these two are assigned to the consumer and producer pointers:

```
rightExpander.producerMessage = (xpander16f*) &xpMsg[0];
rightExpander.consumerMessage = (xpander16f*) &xpMsg[1];
```

The process method now has to:

- check for the presence of an expander on the right;
- if it is present, write to the producer message; and
- flip the producer and consumer pointers.

This is done as follows:

```
bool expanderPresent = (rightExpander.module && rightExpander.module->
  model == modelAPolyXpander);
if (expanderPresent) {
    xpander16f* wrMsg = (xpander16f*)rightExpander.producerMessage;
    for (ch = 0; ch < POLYCHMAX; ch++) {
        wrMsg->outs[ch] = outputs[POLY_SAW_OUT].getVoltage(ch);
    }
    rightExpander.messageFlipRequested = true;
}
```

The first line verifies the presence of a module to the immediate right, and verifies if it is of the modelAPolyXpander kind. If so, a convenience pointer is created of the xpander16f type (remember that rightExpander.producerMessage is a void pointer, and thus we have to cast it). Then we use this pointer to write the individual float values, reading the outputs (the computation of the oscillator outputs has been done already). Finally, we request the flipping, or swapping, of the producer and consumer pointers, which will take place at the next time step.

Now let us get to the Expander. This is developed as any other module, so it consists of two structs, APolyXpander: Module and APolyXpanderWidget: ModuleWidget, and a model, modelAPolyXpander. The module has 16 outputs. To ease your wrists, there is a convenient define, ENUMS, that spares you from writing down all 16 outputs in the OutputIds enum. It simply states that there are 16 outputs as follows:

```
enum OutputIds {
    ENUMS(SEPARATE_OUTS,16),
    NUM_OUTPUTS,
};
```

We will see how to address each one of them. To avoid putting each one of them in the ModuleWidget, we can iterate as follows:

```
for (int i = 0; i < 7; i++) {
    addOutput(createOutputCentered<PJ301MPort>(mm2px(Vec(6.7,
        37+i*11)), module, APolyXpander::SEPARATE_OUTS + i));
}

for (int i = 8; i < 15; i++) {
    addOutput(createOutputCentered<PJ301MPort>(mm2px(Vec(18.2,
        37+(i-8)*11)), module, APolyXpander::SEPARATE_OUTS + i));
}
```

This creates two columns of eight ports each. As you can see, each output is addressed using SEPARATE_OUTS plus an index.

The process method has nothing to do but:

- check for the presence of an APolyDPWOsc module to the left; and
- if it is present, read its data and write the values to the outputs.

```
    bool parentConnected = leftExpander.module && leftExpander.module-
        >model == modelAPolyDPWOsc;
if (parentConnected) {
    xpander16f* rdMsg = (xpander16f*)leftExpander.module->rightExpander.
        consumerMessage;
            for (int ch = 0; ch < POLYCHMAX; ch++) {
                outputs[SEPARATE_OUTS+ch].setVoltage(rdMsg->outs[ch]);
            }
} else {
        for (int ch = 0; ch < POLYCHMAX; ch++) {
                outputs[SEPARATE_OUTS+ch].setVoltage(0.f);
        }
    }
}
```

As you can see, the first line checks if the pointer to the module on the left is not null and whether it is of the APolyDPWOsc type. Then, if the parent is present, it looks for its consumerMessage buffer,[2] sets a convenience pointer to it, and starts copying its values to each individual output of the expander. If the parent module is not present all, outputs are set to zero. The mechanism is depicted graphically in Figure 10.6.

If you want to investigate how the system works, you can start a debug session. By printing the memory address to which producerMessage and consumerMessage point, you will notice that they are swapped at each time step.

Before concluding, we should clarify that the exchange mechanism can be bidirectional. In order to do so, two additional buffers can be allocated in the expander from pushing data to the parent, symmetrically to what has been described so far. Furthermore, to accommodate data from the expander, the data structure can be increased, adding any sort of variables and data types. If, for instance, knobs or buttons are added to the expander to affect the behavior of the DPW oscillator, this data needs to be written back to the oscillator. A larger data structure should supersede xpander16f. Let us define it as:

```
struct xpanderPacket {
    float outs[16];
    bool reset;
};
```

Not all values are written or read at all times: the oscillator module would be responsible for reading the Boolean variable and writing the 16 float values, while the expander module would be responsible for writing reset and reading the 16 float values.

10.4 Adding Factory Presets

If you are a musician or a synth enthusiast, you know very well that you are likely trying the machine presets when you have the chance to try new digital instruments. Good presets can therefore give a very good first impression. As a developer, you must be aware of this. Providing some factory presets for your modules can be a nice idea to showcase some sweet spots (e.g. in the case of a filter, a setting that sounds nice, or in the case of a sequencer, a nice tune). Alternatively, you can prepare the module to serve a specific purpose (e.g. you could provide two settings for a random generator, one for unbiased white noise and the other for random positive CVs).

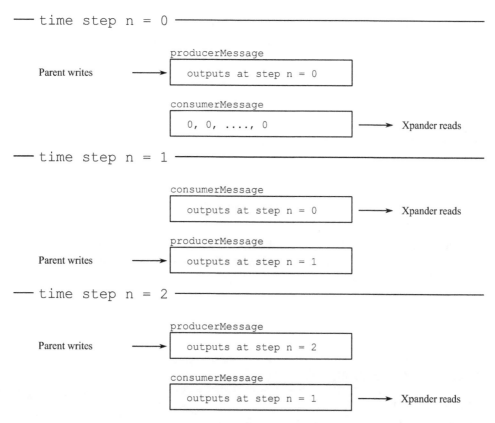

Figure 10.6: The message exchange mechanism used to send data from the parent module to the expander. Two twin data structures are allocated in the parent module. Two pointers, `producerMessage` and `consumerMessage`, point to one such structure each. At each time step, the parent writes to `producerMessage` and the expander reads from `consumerMessage`. At the beginning of each time step, the `producerMessage` and `consumerMessage` pointers are swapped. This introduces a one-sample latency. Please note that for bidirectional data exchange, two other buffers are required. These can be allocated in the expander module and treated similarly.

Providing factory presets is very easy. It is sufficient to follow these instructions:

- Save .vcvm presets by opening the context menu of the module → Presets → Save as.
- Move these into the following directory: Rack/plugins/<myPlugin>/presets/<moduleSlug>/

That's it.

10.5 *Storing and Recalling Module Information through JSON*

Rack allows you to store modules information in JSON format so that any parameter is recalled when a patch is loaded. This is handled automatically by the software for standard information such as the position and wiring of the modules and the status of knobs and buttons. However, you can override the JSON write and read methods to store additional data.

What is JSON, by the way? JavaScript Object Notation, aka JSON, is a data exchange format that is similar in scope to markup languages such as XML. Despite the name, it is agnostic on the programming language, and by now it has libraries for reading and writing JSON in all the most popular programming languages. It is human-readable and easy to parse and is widely used in all fields of ICT. Its main components are objects or tuples of the type:

$$\{key : value\} \tag{10.5}$$

where the key is a string; or arrays of objects such as:

$$[obj_1, obj_2, \ldots] \tag{10.6}$$

where objects can be of any kind: string, integer or real number, another key:value object, another array, a Boolean value, and so on.

To understand this practically, let us examine the content of a VCV file, opening it with a text editor. We start opening VCV Rack with an empty rack and put only one of our modules (e.g. ABC ALinADSR). The file content will be something like:

```
{
  "version": "1.dev",
  "modules": [
   {
    "plugin": "ABC",
    "version": "1.0.0",
    "model": "ALinADSR",
    "params": [
    {
     "id": 0,
     "value": 0.5
    },
    {
     "id": 1,
     "value": 0.5
    {
     "id": 2,
     "value": 0.5
    },
    {
     "id": 3,
     "value": 0.5
    }
   ],
   "id": 22,
   "pos": [
   0,
   0
   ]
  }
  ],
  "cables": []
}
```

This is standard JSON. As you can see, the content is easily understood by visual inspection. In summary, we have a key:value object where "version" and "value"[3] are specified, an array of modules, in this case only ALinADSR, with position (0,0) in the rack, and with the parameters all set to 0.5. Finally, there is an empty cables array, meaning that we did not connect any input/output port.

This is sufficient to store and recall most modules in the Rack community. The operation of storing module information in JSON and recalling from JSON is done by two methods of the Module object:

```
virtual json_t *datatoJson();
virtual void datafromJson(json_t *root);
```

These will take care of storing parameter values and basic module information. However, some special modules may have more data to store. This is the case, for example, for special GUI elements that you design and need the content to be saved explicitly because Rack, by default, knows how to save parameter values. Similarly, you can store the values of custom context menu icons, such as the DPW order in the ADPWOsc module.

Rack employs *jansson*, a C library to store and recall JSON data. This library is part of Rack dependencies, and is downloaded when you execute the make dep command while preparing your setup (see Section 4.4.4). The following function from jansson is used to store a new line:

```
int json_object_set_new(json_t *json, const char *key, json_t *value)
```

where the JSON content is given as the first argument, and the {key, value} tuple is given as second and third arguments, respectively.

To find a value corresponding to a key in a JSON content, you can use the following function from jansson:

```
json_t *json_object_get(const json_t *json, const char *key)
```

where the first argument is the JSON content and the key is the second argument. The function returns a pointer to a JSON element, which contains a value. Please note that it may also return a NULL pointer if the key is not found, so be aware that you need to check it. The pointer can be parsed to get the proper value and store it in one of the variables of your module. For a float value, for example, you can recall the value and store it into a variable as follows:

```
json_t * valueToRecall = json_object_get(root, "myKey");
if (valueToRecall) {
   myFloatVar = json_number_value(valueToRecall);
}
```

10.5.1 Storing and Recalling JSON Data for AModalGUI

We will provide a small example of JSON storage by overriding the toJson and fromJson methods of the AModalGUI class to store the *x* coordinate of the falling mass. When the patch is saved, the last *x* coordinate of the mass will be saved among other parameters. We will also save the number of active oscillators that can be set from the additional context menu items introduced in the related section.

The key strings will be "hitPoint" and "nActiveOsc" (the same as the variables names, for simplicity). The values are float and integer, respectively. We need to declare that we will be overriding the base class methods, so we add the following lines to the declaration of AModalGUI:

```
json_t *dataToJson() override;
void dataFromJson(json_t *rootJ) override;
```

Let us look at the dataToJson implementation first:

```
json_t *AModalGUI::dataToJson() {

    json_t *rootJ = json_object();
    json_object_set_new(rootJ, JSON_NOSC_KEY, json_integer(nActiveOsc));
    json_object_set_new(rootJ, JSON_XCOORD_KEY, json_real(hitPoint));
    return rootJ;
}
```

As you can see, we first instantiate a json_t object. We make use of the json_object_set_new method to set the key and value pairs in the json root object. The key must be a const char *, so passing a constant char array as "hitPoint" is fine, but for simplicity the key strings are defined by JSON_NOSC_KEY and JSON_XCOORD_KEY macros. The value object must be casted to a json_t * type. This is done through convenience functions such as json_real and json_integer, which take a real (double or float) and an integer value, respectively, and return the pointer to json_t.

This function saves data to JSON. Now, to restore this data back during loading, we can do the following.

```
void AModalGUI::dataFromJson(json_t *rootJ) {
    json_t *nOscJ = json_object_get(rootJ, JSON_NOSC_KEY);
    if (nOscJ) {
        nActiveOsc = json_integer_value(nOscJ);
    }
    json_t *xcoorJ = json_object_get(rootJ, JSON_XCOORD_KEY);
    if (xcoorJ) {
        hitPoint = json_number_value(xcoorJ);
        hitPointChanged = true;
    }
}
```

We scan the root for the JSON_NOSC_KEY and JSON_XCOORD_KEY strings using the json_object_get. This function returns a NULL pointer if it does not find the key, so a check is performed before converting it into a numeric value with the json_number_value function and json_integer_value, respectively. The converted value is stored directly into the variable of interest. Special care must be taken for the hit point variable. This belongs to the module struct. However, we need to change the massX variable in the HammDisplay widget in order to really change the position of the mass. A mechanism to propagate this information from the module to the widget follows: we add a Boolean value to the module (defaulting to false) that tells whether the hit point has been changed in the module. We set this to true after reading the new value from JSON. In the HammDisplay draw method, we add the following:

```
if (module->hitPointChanged) {
        massX = (module->hitPoint/DAMP_SLOPE_MAX + 0.5f)*MASS_BOX_W;
        module->hitPointChanged = false;
}
```

This ensures that if hitPoint has been changed from the module, we read it and set massX consequently. We also set hitPointChanged to false, avoiding having to enter this if clause again. Please note that the conversion of the hitPoint into massX is the reverse of what is done in the impact method of AModalGUI:

```
hitPoint = ((x/MASS_BOX_W) - 0.5) * DAMP_SLOPE_MAX;
```

Finally, we can test the code we just added. Open an empty Rack patch, place AModalGUI in it, and click on a position to the right so that the mass falls down. Save the patch and close Rack. Launch Rack and open the patch: the ball will fall in the same *x* coordinate when we saved the patch, allowing the same timbre to be available. If we take a look at the .vcv file, we notice that the "hitPoint" and "nActiveOsc" keys are there:

```
"version": "1.dev",
"modules": [
 {
  "plugin": "ABC",
  "version": "1.0.0",
  "model": "AModalGUI",
  "params": [
   {
    "id": 0,
    "value": 1.0
   },
   {
    "id": 1,
    "value": 0.00999999978
   },
   {
    "id": 2,
    "value": 1.0
   },
   {
    "id": 3,
    "value": 0.0
   },
   {
    "id": 4,
    "value": 0.0
   }
  ],
    "data": {
    "nActiveOsc": 1,
    "hitPoint": 0.000701605692
 },
```

```
    "id": 23,
    "pos": [
    0,
    0
    ]
  }
  ],
  "cables": []
}
```

To conclude this section, strings can be also added to a JSON file. The convenience functions make a string into a json_t pointer and then store it:

```
json_t *json_string(const char *value);
```

The conversion functions to get a string from a json_t pointer are reported below. Additionally, we can also get a string length in advance so that we can allocate a char array to store it:

```
const char *json_string_value(const json_t *string);
size_t json_string_length(const json_t *string);
```

10.6 Versioning the Code and Managing a Git Repository

All plugin developers will need to update their code from time to time (e.g. because of new added features, bugfixes, refactoring, graphics changes, etc.). If you are building and maintaining a plugin for yourself only, it is not strictly necessary to keep a version number or to keep a history of the code (although it does help a lot to do so), but if you are providing your code to the community – and we will see how to do that in Section 10.7 – then you need to be tidy and keep your software versioned.

Code versioning is the practice of managing the code with a *software versioning system* and is fundamental to keep a clear history of the developed code, see changes, implement new features using branches, and manage teamwork. Despite the name, it is not strictly connected to assigning a version number to your Rack plugins. You will advance the version number when you reach a milestone that you want to publish and share outside (see Section 10.7), but with a versioning system, each time you make changes that you want to keep track of, the versioning software will save that version and assign it a number and a comment of yours. The term for this operation is *commit*. I usually commit my code at the end of the day if it compiles and I am happy with the introduced improvements, or I commit multiple times in a day, once for each issue I fix, or once for all issues if they were related.

At some points, branching will come in useful to develop new experimental features while maintaining the codebase deliverable to the users at any time. Think, for example, of the following situation. You have released version 1.0.0 of your plugin. You are preparing version 1.1.0 by rewriting a module entirely to improve its efficiency. A bug is found in the released version that caters the attention of several users. You want to fix it as soon as possible without discarding your work on 1.1.0 (which at the moment looks like open-heart surgery). What you need to do is to branch your code when you start to work on 1.1.0. This will keep the 1.0.0 version of the code available on the main branch. You will be able to switch between the two at any time, being able to fix 1.0.0 and continue developing 1.1.0. After the fix, you will supply version 1.0.1 and will get to finish 1.1.0. After that, you'll want to merge the 1.1.0 branch with the main version.

This is just an example of how useful versioning is. If you are inexperienced with developing and versioning, you will find a few words on the topic here, leaving most of the subject to more specific textbooks.

Several software versioning systems are used nowadays, but probably *git* is the most popular one at the time of writing. Git is also the versioning system employed by VCV Rack, which is well integrated with it. You already learned to use git commands for cloning the Rack repository. Now we see how to manage our repository:

1. Create a GitHub account. There is a plethora of free git repository, but this one is the same used by VCV. Once you have a GitHub account, you will also be able to create tickets in the Rack issue tracker.
2. Create a new (empty) repository from your home page at github.com (e.g. called myPlugin1).
3. Initialize your local repository. Go to an empty folder where you will be fiddling with code (e.g. Documents/sandbox/Rack/plugins/myPlugin1/) and type:

   ```
   git init
   ```

4. Create a README.md file. This is part of GitHub's best practices and is not mandatory. However, we want to add at least one file to our first commit:

   ```
   echo "# myPlugin1" > README.md
   echo "My first Rack plugin" >> README.md
   git add README.md
   git commit -m "initial commit"
   ```

5. Link your local repository to the remote one we just created on GitHub:

   ```
   git remote add origin https://github.com/yourgithubname/myPlugin1.git
   ```

6. Push the locally committed content on the remote repository:

   ```
   git push -u origin master
   ```

Now you can check online that the readme file has been pushed, heading your web browser to https://github.com/yourgithubname/myPlugin1.From now on, if you work alone on the project, you can:

- add existing files to the list of versioned files with the `git add` command;
- create a "commit" (i.e. a small milestone of your project) that will be visible forever on the local repository, with `git commit`; and
- push the local commits to the remote repository so that they will be available to others, with `git push`.

If you are not familiar with git, a nice alternative to using it from the command line is a git client. A good cross-platform choice is *SmartGit*, which is free for non-commercial projects.

This is especially useful for viewing the history, merging your work with others, and managing tags. Tags are useful to keep some of the commits bookmarked with a version number. If, for example, you want to allow a collaborator or an external user to download a specific version of your code, you can tag it with a string such as "v1.2.3" or similar. Please remember that this may also be useful to keep track of the compatibility with Rack versions (e.g. jumping from v1.2.3 to v2.2.3 whenever Rack will switch from v1 to v2).

Please note that you don't need to synchronize to a remote repository; a nice feature of git is that it can also work locally. However, having a remote repository backs you up in case your PC has any hardware or software faults.

Finally, if you sign up for a free GitHub account, your repository will be public. If you want to keep your project closed-source, you need to pay or sign up to some other repository hosting service that leaves you the faculty to make your repository private.

10.7 Getting into the Third-Party Plugin List

As an autonomous developer, you may want to be visible in the third-party plugin list, allowing Rack users to download your plugins and donate. The third-party plugins database is open to anyone willing to share their plugins and their source code, provided that the plugins are not malicious and do not violate any intellectual property. The process to enter the third-party plugin list is quite easy. In the early steps of VCV Rack, plugin developers had to handle the build process for all three platforms themselves and update a JSON file used by Rack to track down the download paths and other information. Nowadays, life is easier thanks to a valuable team of community members. Before asking the Rack team to include your plugins in the list, ensure that:

- the plugin slug is definitive (you will not be able to change it);
- you decided which licensing option to use (GPL, Creative Commons, proprietary, etc.); and
- your plugins compile correctly and are free of bugs (that you are aware of).

These are the steps to follow:

1. Open one thread in the tracker issue of the community GitHub repository (https://github.com/ VCVRack/library/issues) with a title equal to your plugin slug (or multiple slugs, comma-sep- arated, if you have more than one plugin). This thread will be the main channel for communicating with the Rack community members.
2. Provide the URL of your source code to the Rack team by posting a comment in your plugin thread. The Rack team will take care of this and write back if they have issues with your plugin or when they are ready to take you into the third-party plugin list.

You will find up-to-date information at https://github.com/VCVRack/library#for-plugin-developers.

At this point, all users will have access to your plugins via the plugin manager. Whenever you update your modules and want to push the updates to Rack users, do as follows:

1. Increment the "version" field in your plugin.json file.
2. Push a commit to your repository.
3. Post a comment in your plugin's thread with the new version and the commit hash (given by git log or git rev-parse HEAD).

The Rack team will write back to you when ready or if an issue is encountered.

The Rack team will get all the necessary information from the JSON manifest file in your github repository. They will compile the code for all platforms and get back to you if they encounter issues.

10.8 Making a Revenue out of Coding

While Rack and most Rack plugins are available for free and with open-source code, keeping up such awesome work requires resources: food and shelter are not for free. There are therefore ways to get paid for the job of developing and maintaining the code. We shall see what VCV Rack has to offer to the talented developer:

- The VCV official store for selling plugins.
- Putting a donation link in the plugin manager.
- Selling so-called "blank panels" through your e-commerce web page.

Clearly, selling through the store makes sense only if a developer does not release the source code of the plugins he or she intends to be paid for. On the other hand, you can release your code openly but accept donations. These may be received from anyone at any time, most probably by happy users that have downloaded a plugin and are glad to support the developer or that used it in a professional job and want to acknowledge the developer for the quality of his or her plugin with a small contribution. We describe all methods below.

10.8.1 The VCV Store

Selling plugins through the VCV store is not for everyone. The quality of the plugins sold through the VCV store is rather high as they are somewhat featured products, together with the official plugins. If you think you are experienced enough and you have a good collection of modules, you can go that way and contact VCV directly at their email address (currently contact@vcvrack.com). You will be contacted back with all the required information.

10.8.2 Donations

Whether you are a newbie wanting to give it a try or you are a free-software activist accepting no compromise in releasing code to the community, the donation is an option for making a small revenue from your modules. Donations help developers maintain the code, keeping it up to date, cleaning it from bugs, improving its functionalities, and expanding the modules collection. The VCV Rack plugin manager offers the option to add a "Donate" link that allows users to send you an arbitrary amount of money through platforms such as PayPal.

The procedure to add the Donate link to the VCV store is pretty straightforward:

1. Sign up for an account on a money transfer platform such as PayPal. This should give you a link that anyone can use to send you money.
2. Add the link to your plugin manifest (see Section 10.7) under the item "donation."
3. Send a pull request after you have added the manifest to get the plugin manager updated.

Do not be upset if users are not sending you any donations! It is up to them to decide. You can try raising the bar or make your plugins more easily distinguishable from others so people associate your brand with the quality of your modules.

10.8.3 Blank Panels

Another option is to build up your own e-commerce web page where you sell so-called blank panels. The Rack community has developed this peculiar habit of creating blank panels, to "protect your rack from virtual dust" (in the words of Autodafe) and to get economic support. Blank panels are just "useless" modules without any function that users will buy to give a donation to a developer for his or her good work with other plugins. A blank panel is very simple to create!

A blank panel is typically composed of the SVG background (possibly showing your logo or developer nickname), the module widget subclass, and an empty model.

The plugin may just contain the blank panel. There is no specific rule for selling a blank panel, but in general developers sell them through their e-commerce website at a reasonable price that is intended as a small pledge for the other modules by the same author. Blank panels are also useful for advertising as you can boast your logo by placing your blank panel when showing off a patch using your modules.

Notes

1 If allocation is done dynamically, these need be deleted in the module destructor.
2 If dynamic allocation would be used, a check to avoid dereferencing a NULL pointer would be necessary.
3 The version field is empty if you are running from a version of Rack compiled from git.

CHAPTER 11

After Reading This Book

The first aim of this book is to leverage the fun and simplicity of VCV Rack to help the reader get involved with programming, digital signal processing, and music experimentation. This book is thus not only a quick guide to Rack, but also a bridge toward in-depth learning of many other subjects. With this in mind, some suggestions for further reading are left to the reader, in the hope that his or her next read will be a deeper and more complex book on one of the suggested topics. Other suggestions are also provided for those who prefer to play with electronics.
Have fun!

11.1 Forking the ABC Collection

The ABC plugin provides a few essential modules that I found useful to gradually introduce the reader to basic concepts of musical signal processing. However, there is much more to experiment with. First of all, the modules only provide basic functionalities to avoid introducing unnecessary complexity that would deviate the reader from the main point of the discussion. As an example, a real-world sequencer would have many more sequencing options, such as a swing control, tempo signature control, sequence save and recall, auxiliary inputs and outputs, and maybe an internal quantizer to play notes on the equal temperament.

Now that you have learned the basics, you can fork the ABC project and make your own experiments. Some examples to enhance the modules functionalities are provided in the exercises. A few more suggestions for creating new modules that could complement the ones already provided by ABC are:

- a sample and hold (S&H) module;
- a generator of constant CV values;
- an LFO module;
- a VCA module;
- a delay module;
- a chord generator;
- a sample reader or a wavetable oscillator;
- a drum sequencer;
- a linear-FM synthesizer; and
- basic physical modeling modules to interconnect.

You can take inspiration from the huge collection of plugins available from the plugin manager or invent something new.

11.2 Learn More about Virtual Analog Algorithms

There are many resources, by now, on electronic circuits modeling by virtual analog algorithms.

For what concerns basic alias-free oscillator algorithms, you should go on reading about BLEP and related methods. In this book, we covered some basics of oscillators with suppressed aliasing, giving priority to the intuitive DPW method. If you are interested in more oscillator algorithms, you may start reading the tutorial article (Välimäki and Huovilainen, 2007). A family of anti-aliasing algorithms that you need to learn about is the quasi-band-limited algorithms, starting with BLIT (Stilson and Smith, 1996) and further expanded with the BLEP (Brandt, 2001) and BLAMP techniques (Esqueda et al., 2016). Please note that in Välimäki et al. (2010a), the DPW technique is found to be strictly related to the BLIT technique. The minBLEP technique is implemented in include/dsp/minblep.hpp and src/dsp/minblep.cpp so you can implement a minBLEP oscillator starting from it after you are acquainted with its foundations.

A conceptually simpler implementation of oscillators is based on wavetables (Massie, 1998). These require frequent access to the memory and accurate interpolation. The waveshape can be morphed by mixing different wavetables, and a lot of great digital synthesizers from the 1990s on used this technique (and many still do). Aliasing is not an issue with these oscillators as long as you follow some precautions, and once you have set up a working wavetable oscillator it will be ready to take any wavetable you feed to it.

Many other filter topologies not covered here can be studied in Pirkle (2014). The state-variable filter can include nonlinearities. Other filter topologies, including Sallen-Key and the Moog ladder filter, are found on many hardware and software synthesizers. For those who are interested in the discrete-time modeling of analog filter topologies, a fundamental reference is Zavalishin (2018).

More generally, any electronic circuit can now be transposed to the digital domain thanks to the advances in the field. One approach for solving the differential equation of a circuit is the wave method (W-method) (Bilbao, 2004; Sarti and De Sanctis, 2009; de Sanctis and Sarti, 2010), also known as the wave-digital method. Other methods are based on Kirchoff variables, and are hence named Kirchoff methods (K-methods). These were first introduced in Borin et al. (2000), where they are shown to extend previous techniques for delay-free loop elimination (Härmä, 1998) to any sort of nonlinear circuits. Finally, Port-Hamiltonian approaches have recently been proposed in the audio field (Falaize-Skrzek and Hélie, 2013), drawing from the Hamiltonian formulation of systems (Maschke et al., 1992). All these methods and others are resumed in D'Angelo (2014).

Many recent works from Julian Parker discuss West Coast modules such as wavefolders and low-pass gates (Esqueda et al., 2017; Gormond et al., 2018). Other interesting analog effects to emulate are spring reverbs (Välimäki et al., 2010b; Parker, 2011) and tape delays (Zavalishin, 2018). Keep track of the DAFx conferences, held yearly, for updates on the virtual analog field. Their proceedings are available online.

11.3 Experiment with Novel Digital Algorithms

There are a lot of interesting and novel algorithms that could be great to have implemented on a virtual modular synthesizer. First things first, traditional DSP techniques should be the object of thorough study. The classic digital audio effects reference by Udo Zölzer covers all sorts of effects (Zölzer, 2011), while another book by Will Pirkle explains how to implement them in C++ (Pirkle,

2019). Three detailed papers (old but gold) from Jon Dattorro covering reverberation, filtering, modulation effects, and pseudo-noise generation that may be interesting to the reader can also be found (Dattorro, 1997a, 1997b, 2002).

Besides the traditional approaches, interesting novel variations of morphing and hybridization are obtained using non-negative matrix factorization (NMF) (Burred, 2014), the discrete wavelet transform (DWT) (Gabrielli and Squartini, 2012), and live convolution (Brandtsegg et al., 2018). Deep learning is the newest approach for sound synthesis (van den Oord et al., 2016; Gabrielli et al., 2018; Mor et al., 2018; Boilard et al., 2019). At the moment, it is not computationally feasible to fit these algorithms in real-time music processing platforms, but in the years to come this may change. Tones can be also generated offline by these algorithms and be processed in real time, as is done for the NSynthsuper project, which mixes tones generated by the Nsynth neural synthesizer (Engel et al., 2017). Machine learning is also a valuable tool for generative music. Generating musical events is not so demanding in computational terms, and thus a generative model based on neural networks or other machine learning techniques is feasible. A thorough review of computational intelligence techniques for generative music is provided in Liu and Ting (2017). Cellular automata have been one of the preferred choices for generative music for decades (Burraston and Edmonds, 2005). Currently, a lot of tools are released openly by Google AI researchers taking part in the Magenta project (https://magenta.tensorflow.org/), mainly based on JavaScript and Python.

Novel types of oscillators can be designed by discretizing chaotic nonlinear dynamical systems. Examples are the Chua-Felderhoff circuit (Borin et al., 2000) and the Lotka-Volterra biological population model (Fontana, 2012). Putting these into a novel oscillator module may be an interesting idea. Other computationally efficient oscillators that can produce crazy and unpredictable sounds are those based on feedback amplitude modulation (Kleimola et al., 2011b; Timoney et al., 2014). Vector phaseshaping is another interesting abstract form of synthesis (Kleimola et al., 2011a).

11.4 Build a Hybrid Hardware/Software Modular Rack

Computing devices, and laptops in particular, are nowadays pervasive in musical practice. There is, however, an increasing reluctance in the use of the laptop as a live performance tool and an increasing need for a more physical interaction for music production. This is one reason why modular synthesizers are getting more and more popular. However, with Rack, you still have to deal with a laptop. But there are chances of getting the experience to a higher level by building Rack into a modular case and integrating it with hardware modules.

Figure 11.1 shows a proof of concept of such a system. It is based on a small form factor computer (Mini-ITX) with a quite inexpensive CPU (Intel 7th gen i3), a 14″ screen, a resistive digitizer, a robust case, and a Eurorack device acting as a sound card.

Figure 11.2 shows the conceptual scheme for the design of such a system. The system is based on a 64-bit x86 CPU hosted by a motherboard (at the time of writing, the Mini-ITX format offers the best trade-off between cost, size, and performance). The motherboard is fed by a power supply of possibly small form factor to reduce the size of the case. An LCD screen is connected to the board through a screen controller via a VGA, DVI, or HDMI connector. The screen is touch-sensitive thanks to a resistive or capacitive screen digitizer that sends touch information through a controller via USB. Integrated touchscreens exist, but you can also spare some old laptop LCD screen and just

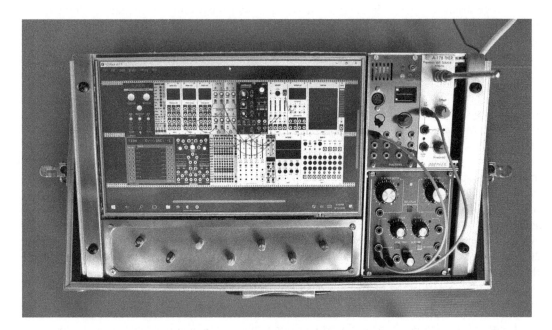

Figure 11.1: A proof of concept of a virtual modular Eurorack synthesizer with touchscreen and an embedded PC. A digital Eurorack module acts as a DC-coupled signal interface (module on the bottom row). Two Eurorack modules act as input devices: a proximity antenna (top row, right) and a brainwave reader (to be connected to a Bluetooth EEG headset). Additional potentiometers allow expressive control of the performance (below the screen).

buy the digitizer. The environment will be grateful. I suggest buying a digitizer that comes mounted on robust glass to make it live gig-proof. Any sound card can be mounted, of course, but the best way to integrate this setup with Eurorack modules is to use a DC-coupled interface, able to read and generate control voltage signals, otherwise you will not be able to send and receive CV signals, only audio signals. Finally, the project can be complemented with Eurorack modules, either mounted on board or mounted on a separate case, and eventually a set of knobs and sliders that are read by a microcontroller. Arduino boards and the like are a good option and are easy to program. These can read the voltage across the potentiometers and send MIDI data accordingly through a USB cable to the board. This provides additional knobs that will improve the live performance.

During the installation of the PC you will need an external keyboard and mouse, but once the touchscreen is up and working you can use the on-screen keyboard provided by the operating system. Ubuntu provides a great touch experience from version 18.04, but Windows 10 also has a nice interface (tablet mode).

At the time of writing, the use of Rack with a touchscreen is not officially supported, but it is much more engaging to use it with a touchscreen rather than a mouse. One fundamental option must be tweaked to allow moving knobs with a touchscreen.

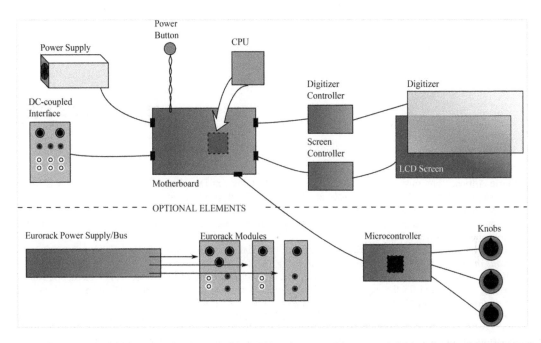

Figure 11.2: **The architecture of the proposed hybrid hardware/software modular synthesizer based on VCV Rack. The optional elements include Eurorack modules and a microcontroller to expand the capabilities of the system by adding additional control elements (e.g. knobs) that send MIDI messages to the computer via MIDI-USB link.**

In the Rack folder, open settings.json with a text editor and change the allowCursorLock to false:

"allowCursorLock": false

Without this change, there will be no chance of moving a knob with a touch of your finger.

As the one proposed here is just a proof of concept, I expect enthusiasts to build even better solutions in the future.

Bibliography

Jean-Marie Adrien, "The missing link: Modal synthesis." In Giovanni De Poli, Aldo Piccialli, and Curtis Roads, Eds., *Representations of Musical Signals*, pp. 269–297. MIT Press, Cambridge, MA, 1991.

Federico Avanzini, Stefania Serafin, and Davide Rocchesso, "Interactive simulation of rigid body interaction with friction-induced sound generation." *IEEE Transactions on Audio, Speech, and Language Processing* 13.6 (November 2005): 1073–1081.

Balázs Bank, Stefano Zambon, and Federico Fontana, "A modal-based real-time piano synthesizer." *IEEE Transactions on Audio, Speech, and Language Processing* 18.4 (2010): 809–821.

Stefan Bilbao, *Wave and Scattering Methods for Numerical Simulation*. John Wiley & Sons, Hoboken, NJ, 2004.

Harald Bode, "A new tool for the exploration of unknown electronic music instrument performances." *Journal of the Audio Engineering Society* 9 (October 1961): 264.

Harald Bode, "History of electronic sound modification." *Journal of the Audio Engineering Society* 32.10 (1984): 730–739.

Jonathan Boilard, Philippe Gournay, and Roch Lefebvre, "A literature review of WaveNet: Theory, application, and optimization." In *Audio Engineering Society Convention 146*, Dublin, Ireland, 2019.

Gianpaolo Borin, Giovanni De Poli, and Davide Rocchesso, "Elimination of delay-free loops in discrete-time models of nonlinear acoustic systems", *IEEE Transactions on Speech and Audio Processing* 8.5 (September 2000): 597–605.

E. Brandt, "Hard sync without aliasing." In *Proceedings of International Computer Music Conference (ICMC)*, Havana, Cuba, 2001, pp. 365–368.

Øyvind Brandtsegg, Sigurd Saue, and Victor Lazzarini, "Live convolution with time-varying filters." *Applied Sciences* 8.1 (2018).

Dave Burraston and Ernest Edmonds, "Cellular automata in generative electronic music and sonic art: A historical and technical review." *Digital Creativity* 16.3 (2005): 165–185.

Juan José Burred, "A framework for music analysis/resynthesis based on matrix factorization." In *The Proceedings of the International Computer Music Conference*, Athens, Greece, 2014, pp. 1320–1325.

Hal Chamberlin, *Musical Applications of Microprocessors*. Hayden Book Company, Indianapolis, IN, 1985.

John Chowning, "The synthesis of complex audio spectra by means of frequency modulation." *Journal of the Audio Engineering Society* 21.7 (1973): 526–534.

Perry R. Cook, "Physically informed sonic modeling (PhISM): Synthesis of percussive sounds." *Computer Music Journal* 21.3 (1997): 38–49.

James W. Cooley and John W. Tukey, "An algorithm for the machine calculation of complex Fourier series." *Mathematics of Computation* 19.90 (1965): 297–301.

Stefano D'Angelo, *Virtual Analog Modeling of Nonlinear Musical Circuits*. PhD thesis, Aalto University, Espoo, Finland, November 2014.

Stefano D'Angelo, Leonardo Gabrielli, and Luca Turchet, "Fast approximation of the Lambert W Function for virtual analog modeling." In *Proc. DAFx*, Birmingham, UK, 2019.

Jon Dattorro, "Effect design – Part 1: Reverberator and other filters." *Journal of the Audio Engineering Society* 45.9 (1997a): 660–684.

Jon Dattorro, "Effect design – Part 2: Delay-line modulation and chorus." *Journal of the Audio Engineering Society* 45.10 (1997b): 764–788.

Jon Dattorro, "Effect design – Part 3: Oscillators – sinusoidal and pseudonoise." *Journal of the Audio Engineering Society* 50.3 (2002): 115–146.

Giovanni de Sanctis and Augusto Sarti, "Virtual analog modeling in the wave-digital domain." *IEEE Transactions on Audio, Speech, and Language Processing* 18.4 (May 2010): 715–727.

Paolo Donati and Ettore Paccetti, *C'erano una volta nove oscillatori: lo studio di Fonologia della RAI di Milano nello sviluppo della nuova musica in Italia*. RAI Teche, Rome, 2002.

Jesse Engel, Cinjon Resnick, Adam Roberts, Sander Dieleman, Mohammad Norouzi, Douglas Eck, and Karen Simonyan, "Neural audio synthesis of musical notes with WaveNet autoencoders." In *Proceeding of the 34th International Conference on Machine Learning (ICML17) – Volume 70*, Sydney, NSW, Australia, 2017, pp. 1068–1077.

Fabián Esqueda, Henri Pontynen, Julian Parker, and Stefan Bilbao, "Virtual analog model of the Lockhart wavefolder." In *Proceedings of the Sound and Music Computing Conference (SMC)*, Espoo, Finland, 2017, pp. 336–342.

Fabián Esqueda, Vesa Välimäki, and Stefan Bilbao, "Rounding corners with BLAMP." In *Proceedings of the 19th International Conference on Digital Audio Effects (DAFx-16)*, Brno, Czech Republic, 5–9 September 2016, pp. 121–128.

Antoine Falaize-Skrzek and Thomas Hélie, "Simulation of an analog circuit of a wah pedal: A Port-Hamiltonian approach." In *Proceedings of 135th AES Convention*, New York, October 2013.

Agner Fog, *Optimization Manuals*, Chapter 4: "Instruction tables." Technical report, 2018, available online at: www.agner.org/optimize/instruction_tables.pdf (last accessed 4 November 2018).

Federico Fontana, "Interactive sound synthesis by the Lotka-Volterra population model." In *Proceedings of the 19th Colloquium on Music Informatics*, Trieste, Italy, 2012.

Gene F. Franklin, J. David Powell, and Abbas Emami-Naeini, *Feedback Control of Dynamic Systems*, 7th Edition, Pearson, London, 2015.

Dennis Gabor, "Theory of communication – Part 1: The analysis of information." *The Journal of the Institution of Electrical Engineers* 93.26 (1946): 429–441.

Dennis Gabor, "Acoustical quanta and the theory of hearing." *Nature* 159.4044 (1947): 591–594.

L. Gabrielli, M. Giobbi, S. Squartini, and V. Välimäki, "A nonlinear second-order digital oscillator for virtual acoustic feedback." In *2014 IEEE International Conference on Acoustics, Speech and Signal Processing (ICASSP)*, Florence, 2014, pp. 7485–7489.

Leonardo Gabrielli, Carmine E. Cella, Fabio Vesperini, Diego Droghini, Emanuele Principi, and Stefano Squartini, "Deep learning for timbre modification and transfer: An evaluation study." In *Audio Engineering Society Convention 144*, Milan, Italy, 2018.

Leonardo Gabrielli and Stefano Squartini, "Ibrida: A new DWT-domain sound hybridization tool." In *AES 45th International Conference*, Helsinki, Finland, 2012.

Leonardo Gabrielli, and Stefano Squartini, *Wireless Networked Music Performance*, Springer, New York, 2016.

David Goldberg, "What every computer scientist should know about floating-point arithmetic." *ACM Computing Surveys (CSUR)* 23.1 (1991): 5–48.

Geoffrey Gormond, Fabián Esqueda, Henri Pöntynen, and Julian D. Parker, "Waveshaping with Norton amplifiers: Modeling the Serge triple waveshaper." In *International Conference on Digital Audio Effects*, Aveiro, Portugal, 2018, pp. 288–295.

Aki Härmä, "Implementation of recursive filters having delay free loops." In *Proc. Intl. Conf. Acoust., Speech, and Signal Process. (ICASSP 1998), vol. 3*, Seattle, WA, May 1998, pp. 1261–1264.

David H. Howard and Jamie A.S. Angus, *Acoustics and Psychoacoustics*, 5th Edition, Focal Press, Waltham, MA, 2017.

Jari Kleimola, Victor Lazzarini, Joseph Timoney, and Vesa Välimäki, "Vector phaseshaping synthesis." In *The Proceedings of the 14th Int. Conference on Digital Audio Effects (DAFx-11)*, Paris, France, September 2011a.

Jari Kleimola, Victor Lazzarini, Vesa Välimäki, and Joseph Timoney, "Feedback amplitude modulation synthesis." *EURASIP Journal on Advances in Signal Processing* (2011b).

Donald E. Knuth, "Structured programming with go to statements." *ACM Computing Surveys (CSUR)* 6.4 (1974): 261–301.

Victor Lazzarini, "Supporting an object-oriented approach to unit generator development: The Csound Plugin Opcode Framework." *Applied Sciences* 7.10 (2017).

Claude Lindquist, *Adaptive and Digital Signal Processing*. Steward & Sons, 1989.

Chien-Hung Liu and Chuan-Kang Ting, "Computational intelligence in music composition: A survey". *IEEE Transactions on Emerging Topics in Computational Intelligence* 1.1 (February 2017): 2–15.

Bernhard Maschke, Arjan van der Schaft, and P.C. Breedveld, "An intrinsic Hamiltonian formulation of network dynamics: Non-standard Poisson structures and gyrators." *J. Franklin Institute* 329.5 (September 1992): 923–966.

Dana C. Massie, "Wavetable sampling synthesis." In M. Kahrs and K. Brandenburg, Eds., *Applications of Digital Signal Processing to Audio and Acoustics*, pp. 311–341. Kluwer, Boston, MA, 1998.

Robert A. Moog, "Voltage-controlled electronics music modules." *Journal of the Audio Engineering Society*, 13.3 (July 1965): 200–206.

Noam Mor, Lior Wolf, Adam Polyak, and Yaniv Taigman, "A Universal Music Translation Network. *arXiv* pre-print arXiv:1805.07848 (2018).

Juhan Nam, Vesa Välimäki, Jonathan S. Abel, and Julius O. Smith, "Efficient antialiasing oscillator algorithms using low-order fractional delay filters." *IEEE Transactions on Audio Speech and Language Processing* 18.4 (2010): 773–785.

K.S. Narendra and P.G. Gallman, "An iterative method for the identification of nonlinear systems using a Hammerstein model." *IEEE Transactions on Automatic Control* 11.7 (1966): 546–550.

Maddalena Novati and Ricordi John Dack, Eds., *The Studio Di Fonologia: A Musical Journey.* BMC Ricordi, Milan, 2012.

Harry Nyquist, "Certain topics in telegraph transmission theory." *Transactions of the American Institute of Electrical Engineers* 47.2 (1928): 617–644.

Nelles Oliver, *Nonlinear System Identification: From Classical Approaches to Neural Networks and Fuzzy Models.* Springer, New York, 2001.

Alan V. Oppenheim, and Ronald W. Schafer, *Digital Signal Processing*, 3rd Edition. Prentice Hall, Upper Saddle River, NJ, 2009.

Julian Parker, "Efficient dispersion generation structures for spring reverb emulation." *EURASIP Journal on Advances in Signal Processing* 2011.1 (2011): 646134.

Julian D. Parker, Vadim Zavalishin, and Efflam Le Bivic, "Reducing the aliasing of nonlinear waveshaping using continuous-time convolution." In *Proceedings of Int. Conf. Digital Audio Effects (DAFx-16)*, Brno, Czech Republic, 2016, pp. 137–144.

Jussi Pekonen, Victor Lazzarini, Joseph Timoney, Jari Kleimola, and Vesa Välimäki, "Discrete-time modeling of the Moog Sawtooth oscillator waveform." *EURASIP Journal on Advances in Signal Processing* (2011).

Will Pirkle, *Designing Software Synthesizer Plug-Ins in C++: For RackAFX, VST3, and Audio Units.* Focal Press, Waltham, MA, 2014.

Will Pirkle, *Designing Audio Effect Plug-Ins in C++*, 2nd Edition. Focal Press, Waltham, MA, 2019.

Augusto Sarti and Giovanni De Sanctis, "Systematic methods for the implementation of nonlinear wave digital structures." *IEEE Transactions on Circuits and Systems I, Regular Papers* 56 (2009): 470–472.

Martin Schetzen, *The Volterra and Wiener Theories of Nonlinear Systems.* John Wiley & Sons, Hoboken, NJ, 1980.

Claude E. Shannon, "Communication in the presence of noise." In *Proceedings of the Institute of Radio Engineers, 1949.* Reprinted in: *Proceedings of the IEEE* 86.2, February 1998, pp. 447–457.

Julius O. Smith III and Perry R. Cook, "The second-order digital waveguide oscillator." In *Proceedings of the 1992 International Computer Music Conference*, San Jose, CA, 1992, pp. 150–153.

T. Stilson and Julius O. Smith, "Alias-free digital synthesis of classic analog waveforms." In *Proceedings of International Computer Music Conference*, Hong Kong, China, 1996, pp. 332–335.

Karlheinz Stockhausen, *Texte Zur Musik 1984–1991*, Vol. 8 Dienstag aus Licht; Elektronische Musik, Eds. Christoph von Blumröder. Stockhausen-Verlag, Kürten, 1998.

Joseph Timoney, Jussi Pekonen, Victor Lazzarini, and Vesa Välimäki, "Dynamic signal phase distortion using coefficient-modulated Allpass filters." *Journal of the Audio Engineering Society* 62.9 (September 2014): 596–610.

Vesa Välimäki, "Discrete-time synthesis of the sawtooth waveform with reduced aliasing." *IEEE Signal Processing Letters* 12.3 (March 2005): 214–217.

Vesa Välimäki and Antti Huovilainen, "Antialiasing oscillators in subtractive synthesis." *IEEE Signal Processing Magazine* 24.2 (2007): 116–125.

Vesa Välimäki, Juhan Nam, Julius O. Smith, and Jonathan Abel, "Alias-suppressed oscillators based on differentiated polynomial waveforms." *IEEE Transactions on Audio, Speech, and Language Processing* 18.4 (2010a): 786–798.

Vesa Välimäki, Julian Parker, and Jonathan S. Abel, "Parametric spring reverberation effect." *Journal of the Audio Engineering Society* 58.7/8 (2010b): 547–562.

Aaron van Den Oord, Sander Dieleman, Heiga Zen, Karen Simonyan, Oriol Vinyals, Alex Graves, Nal Kalchbrenner, Andrew Senior, and Koray Kavukcuoglu, *Wavenet: A Generative Model for Raw Audio*. White paper, 2016, available online at: https://arxiv.org/pdf/1609.03499.pdf (last accessed 30 October 2019).

Martin Vicanek, *A New Recursive Quadrature Oscillator*. White paper, 21 October 2015, available online at: www.vicanek.de/articles/QuadOsc.pdf (last accessed 30 March 2019).

N. Wiener, "Response of a nonlinear system to noise." Technical report, Radiation Lab MIT 1942, restricted. Report V-16, no. 129 (112 pp.). Declassified July 1946. Published as rep. no. PB-1-58087, U.S. Department of Commerce, 1942.

Adrian Wills, Thomas B. Schön, Lennart Ljung, and Brett Ninness, "Identification of Hammerstein–Wiener models." *Automatica* 49.1 (January 2013): 70–81.

Duane K. Wise, "The modified Chamberlin and Zölzer filter structures." In *Proceedings of the 9th International Digital Audio Effects Conference (DAFX06)*, Montreal, Canada, 2006, pp. 53–56.

Lior Wolf, Adam Polyak, and Yaniv Taigman, "A universal music translation network." *arXiv* preprint arXiv: 1805.07848 (2018).

Stefano Zambon, Leonardo Gabrielli, and Balazs Bank, "Expressive physical modeling of keyboard instruments: From theory to implementation." In *Audio Engineering Society Convention 134*, Rome, Italy, 2013.

Vadim Zavalishin, *The Art of VA Filter Design*, 2018, available online at: www.native-instruments.com/fileadmin/ni_media/downloads/pdf/VAFilterDesign_2.1.0.pdf (last accessed 28 May 2018).

Vadim Zavalishin and Julian D. Parker, "Efficient emulation of tapelike delay modulation behavior." In *Proceedings of 21st Int. Conf. Digital Audio Effects (DAFx-18)*, Aveiro, Portugal, 2018, pp. 3–10.

Udo Zölzer, *DAFX: Digital Audio Effects*, 2nd Edition, John Wiley & Sons, Hoboken, NJ, 2011.

Index

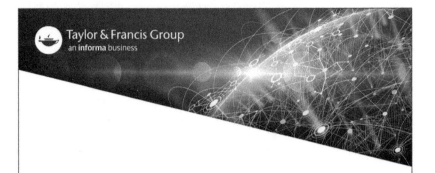